Signals and Communication Technology

For further volumes:
http://www.springer.com/series/4748

Tobias Herbig, Franz Gerl, and
Wolfgang Minker

Self-Learning Speaker Identification

A System for Enhanced Speech Recognition

Springer

Author

Tobias Herbig
Institute of Information Technology
University of Ulm
Albert-Einstein-Allee 43
89081 Ulm
Germany
E-mail: tobias.herbig@uni-ulm.de

Wolfgang Minker
Institute of Information Technology
University of Ulm
Albert-Einstein-Allee 43
89081 Ulm
Germany
E-mail: wolfgang.minker@uni-ulm.de

Dr. Franz Gerl
Harman Becker Automotive
Systems GmbH
Dept. Speech Recognition
Söflinger Str. 100
89077 Ulm
Germany
E-mail: franz.gerl@harman.com

ISBN 978-3-642-19898-4

e-ISBN 978-3-642-19899-1

DOI 10.1007/978-3-642-19899-1

Library of Congress Control Number: 2011927891

Typeset: Scientific Publishing Services Pvt. Ltd., Chennai, India.

Printed on acid-free paper

9 8 7 6 5 4 3 2 1

springer.com

Self-Learning Speaker Identification for Enhanced Speech Recognition

Summary. A self-learning speech controlled system has been developed for unsupervised speaker identification and speech recognition. The benefits of a speech controlled device which identifies its main users by their voice characteristics are obvious: The human-computer interface may be personalized. New ways for interacting with a speech controlled system may be developed to simplify the handling of the device. Furthermore, speech recognition accuracy may be significantly improved if knowledge about the current user can be employed. The speech modeling of a speech recognizer can be improved for particular speakers by adapting the statistical models on speaker specific data. The adaptation scheme presented captures speaker characteristics after a very few utterances and transits smoothly to a highly specific speaker modeling. Speech recognition accuracy may be continuously improved. Therefore it is quite natural to employ speaker adaptation for a fixed number of users. Optimal performance may be achieved when each user attends a supervised enrollment and identifies himself whenever the speaker changes or the system is reset. A more convenient human-computer communication may be achieved if users can be identified by their voices. Whenever a new speaker profile is initialized a fast yet robust information retrieval is required to identify the speaker on successive utterances. Such a scenario presents a unique challenge for speaker identification. It has to perform well on short utterances, e.g. commands, even in adverse environments. Since the speech recognizer has a very detailed knowledge about speech, it seems to be reasonable to employ its speech modeling to improve speaker identification. A unified approach has been developed for simultaneous speech recognition and speaker identification. This combination allows the system to keep long-term adaptation profiles in parallel for a limited group of users. New users shall not be forced to attend an inconvenient and time-consumptive enrollment. Instead new users should be detected in an unsupervised manner while operating the device. New speaker profiles have to be initialized based on the first occurrences of a speaker. Experiments on the evolution of such a system were carried out on a subset of the SPEECON database. The results show that in the long run the system produces adaptation profiles which give continuous improvements in speech recognition and speaker identification rate. A variety of applications may benefit from a system that adapts individually to several users.

Contents

Nomenclature

ADC	Analog to Digital Converter
ASR	Automated Speech Recognition
BIC	Bayesian Information Criterion
CDHMM	Continuous Density HMM
CMS	Cepstral Mean Subtraction
cos	cosine
DCT	Discrete Cosine Transform
DFT	Discrete Fourier Transform
EM	Expectation Maximization
EMAP	Extended Maximum A Posterior
EV	Eigenvoice
FFT	Fast Fourier Transform
GMM	Gaussian Mixture Model
HMM	Hidden Markov Model
Hz	Hertz
ID	Identity
iid	independent and identically distributed
kHz	kilo Hertz
LDA	Linear Discriminant Analysis
LLR	Log Likelihood Ratio
log	logarithm
LR	Lip Radiation
MAP	Maximum A Posteriori
MFCC	Mel Frequency Cepstral Coefficients
min	minute
ML	Maximum Likelihood
MLLR	Maximum Likelihood Linear Regression
MLP	Multilayer Perceptron
MMSE	Minimum Mean Squared Error
NN	Neural Networks
PCA	Principal Component Analysis

RBF	Radial Basis Functions
ROC	Receiver Operator Characteristics
SCHMM	Semi-Continuous HMM
sec	second
SPEECON	Speech-Driven Interfaces for Consumer Devices
SPLICE	Stereo-based Piecewise Linear Compensation for Environments
STFT	Short Time Fourier Transform
UBM	Universal Background Model
VAD	Voice Activity Detection
VQ	Vector Quantization
VT	Vocal Tract
VTLN	Vocal Tract Length Normalization
WA	Word Accuracy
WER	Word Error Rate

1 Introduction

Automatic speech recognition has attracted various research activities since the 1950s. To achieve a high degree of user-friendliness, a natural and easy to use human-computer interface is targeted for technical applications. Since speech is the most important means of interpersonal communication, machines and computers can be operated more conveniently with the help of automated speech recognition and understanding [Furui, 2009].

The history of speech recognition and speaker identification is characterized by steady progress. Whereas in the 1950s the first realizations were mainly built on heuristic approaches, more sophisticated statistical techniques have been established and continuously developed since the 1980s [Furui, 2009]. Today a high level of recognition accuracy has been achieved which allows applications of increasing complexity to be controlled by speech.

Speech recognition has attracted attention for a variety of applications such as office systems, manufacturing, telecommunication, medical reports or infotainment systems in automobiles [Rabiner and Juang, 1993]. Speech recognition can increase labor efficiency in call-centers or for dictation tasks of special occupational groups with extensive documentation duty. For in-car applications both the usability and security can be increased for a wide variety of users. The driver can be supported to safely participate in road traffic and to operate technical devices, e.g. navigation systems or hands-free sets.

Despite significant advances during the last few decades, there still exist some deficiencies that limit the wide-spread application of speech recognition in technical applications [Furui, 2009]. For example, recognition accuracy can be negatively affected by changing environments, speaker variability and natural language input [Junqua, 2000].

The goal of this book is to contribute to a natural human-computer communication. An automatic personalization through a self-learning speech controlled system is targeted. An integrated implementation of speech recognition and speaker identification has been developed which adapts individually to several users. Significant progresses for speech recognition are exemplified for in-car applications.

T. Herbig, F. Gerl, and W. Minker: Self-Learning Speaker Identification, SCT, pp. 1–4.
springerlink.com

1.1 Motivation

Infotainment systems with speech recognition, in general navigation, telephone or music control, typically are not personalized to a single user. The speech signal may be degraded by varying engine, wind and tire noises, or transient events such as passing cars or babble noise. For embedded systems computational efficiency and memory consumption are important design parameters. Nevertheless, a very large vocabulary, e.g. city or street names, needs to be accurately recognized.

Speaker independent speech recognizers are trained on a large set of speakers. Obviously, the trained speech pattern does not fit the voice of each speaker perfectly [Zavaliagkos et al., 1995]. To achieve high recognition rates in large vocabulary dictation systems, the statistical models can be perfectly trained to a particular user [Thelen, 1996]. However, the user is known in this case and the environment in general does not change.

In a system without speaker tracking all information acquired about a particular speaker is lost with each speaker turn. However, often one can expect that a device is only used by a small number of users, e.g. 5 recurring speakers. Therefore, it seems to be natural to employ speaker adaptation separately for several speakers and to integrate speaker identification. In addition to an enhanced speech recognition accuracy by incremental speaker adaptation, the usability can be improved by tracking personal preferences and habits.

A simple implementation would be to impose the user to identify himself whenever the system is initialized or the speaker changes. A more natural human-computer communication will be achieved by identifying the current user in an unsupervised way. No additional intervention of the user should be required.

Automatic personalization through a self-learning speech controlled system is targeted in this book. An integrated implementation comprising speech recognition, speaker identification, detection of new users and speaker adaptation has been developed. Speech recognition is enhanced by individually adapting a speech recognizer to 5-10 recurring users in an unsupervised manner. Several tasks have to be accomplished by such a system:

The various speaker adaptation schemes which have been developed, e.g. by Gauvain and Lee [1994]; Kuhn et al. [2000]; Stern and Lasry [1987], mainly differ in the number of parameters which have to be estimated. In general, a higher number of parameters allows more individual representation of the speaker's characteristics leading to improved speech recognition results. However, such approaches are only successful when a sufficient amount of utterances proportional to the number of adaptation parameters can be reliably attributed to the correct speaker. Especially during the initialization of a new speaker profile, fast speaker adaptation that can converge within a few utterances is essential to provide robust statistical models for efficient speech recognition and reliable speaker identification. When prior knowledge about speaker variability can be used only a few parameters are required to efficiently adapt the statistical models of a speech recognizer. However, the capability to track individual characteristics can be strongly limited. In the long run speaker characteristics have to be optimally captured. Thus, a balanced strategy of fast and individual adaptation has to be developed by combining the respective advantages of both approaches.

Speech recognition has to be extended so that several individually trained speaker profiles can be applied in parallel. Optimal recognition accuracy may be achieved for each user by an efficient profile selection during speech recognition. Recognition accuracy and computational complexity have to be addressed.

Speaker identification is essential to track recurring users across speaker turns to enable optimal long-term adaptation. Speakers have to be identified reliably despite limited adaptation data to guarantee long-term stability and to support the speech recognizer to build up well trained speaker profiles. Unknown speakers have to be detected quickly so that new profiles can be initialized in an unsupervised way without any intervention of the user. An enrollment to train new profiles should be avoided for the sake of a more natural operation of speech controlled devices.

Since applications for embedded systems are examined, the system shall be designed to efficiently retrieve and represent speech and speaker characteristics, especially for real-time computation. Multiple recognitions of speaker identity and spoken text have to be avoided. A unified statistical modeling of speech and speaker related characteristics will be presented as an extension of common speech recognizer technologies.

Beyond the scope of this book, the knowledge about the speaker is not only useful for speech recognition. The human-computer interface may be personalized in different ways:

Feedback of the speaker identity to the speech dialog engine allows special habits and knowledge of the user about the speech controlled device to be taken into consideration. For instance, two operator modes for beginners and more skilled users are possible. Beginners can be offered a comprehensive introduction or assistance concerning the handling of the device. Advanced users can be pointed to more advanced features.

Knowledge about the speaker identity enables the system to prefer or preselect speaker specific parameter sets such as an address book for hands-free telephony or a list of frequently selected destinations for navigation. Confusions of the speech recognizer can be reduced and in doubt the user can be offered a list of reasonable alternatives leading to an increased usability.

Barge-in detection is another application where techniques for speaker identification and speech recognition can be of help as shown by Ittycheriah and Mammone [1999]; Ljolje and Goffin [2007]. Barge-in allows the user to interrupt speech prompts, e.g. originating from a Text-To-Speech (TTS) system, without pressing a push-to-talk button. A more natural and efficient control of automated human-computer interfaces may be achieved.

1.2 Overview

First, the fundamentals relevant for the scope of this book are discussed. Speech production is explained by a simple model. It provides a first overview of speech and speaker variability. The core components of a speech controlled system, namely signal processing and feature extraction, speaker change detection, speaker identification, speech recognition and speaker adaptation, are introduced and discussed in detail. Step by step more sophisticated strategies and statistical models are explained to handle speech and speaker characteristics.

Then a brief survey of more complex approaches and systems concerning the intersection of speaker change detection, speaker identification, speech recognition and speaker adaptation is given. Speech and speaker characteristics can be involved in several ways to identify the current user of a speech controlled system and to understand the content of the spoken phrase. The target system of this book is sketched and the main aspects are emphasized.

The first component of the target system is a speaker adaptation scheme which combines short-term adaptation as well as individual adjustments of the speaker profiles in the long run. This speaker adaptation method is able to capture speaker characteristics after a few utterances and transits smoothly to a highly specific speaker modeling. It is used subsequently to initialize and continuously adapt the speaker profiles of the target system and forms the identification basis of an integrated approach for speaker identification and speech recognition. Remarkable improvements for speech recognition accuracy may be achieved under favorable conditions.

Another important component of the target system is the unified realization of speaker identification and speech recognition which allows a compact statistical representation for both tasks. Speaker identification enables the system to continuously adapt the speaker profiles of its main users. Speech recognition accuracy is continuously improved. Experiments have been conducted to show the benefit for speech recognition and speaker identification.

The basic system can be extended by long-term speaker tracking. Speaker tracking across several utterances significantly increases speaker identification accuracy and ensures long-term stability of the system. The evolution of the adaptive system can be taken into consideration so that a strictly unsupervised system may be achieved. New users may be detected in an unsupervised way.

Finally, a summary and conclusion are given which repeat the main aspects and results. Several future applications and extensions are discussed in an outlook.

2

Fundamentals

The fundamentals of speech production, automatic speech processing and the components of a complete system for self-learning speaker identification and speech recognition are introduced in this chapter.

In the first section discussion starts with speech production to gain insight into speech and the problem of speaker variability. A simple yet effective model is provided to simplify the understanding of the algorithms handling speaker variability and speech characteristics.

Then feature extraction is considered as the inverse process. A technique is introduced which extracts the relevant speech features from the recorded speech signal that can be used for further automated processing.

Based on these features the modules speaker change detection, speaker identification and speech recognition are described. They are essential for a speech controlled system operated in an unsupervised manner. Starting from a basic statistical model, each module introduces an additional extension for a better representation of speech signals motivated by speech production theory.

Finally, two competing approaches are presented to handle unseen situations such as speaker variability or acoustic environments. Speaker adaptation allows modifying the statistical models for speech and speaker characteristics whereas feature vector normalization or enhancement compensates mismatches on a feature level. A continuous improvement of the overall performance is targeted.

Based on the fundamentals, several realizations of complete systems known from literature are presented in the next chapter.

2.1 Speech Production

Fant's model [Fant, 1960] defines speech production as a source-filter model[1]. The air stream originating from the lungs flows through the vocal cords and

[1] The subsequent description of speech production follows O'Shaughnessy [2000] if not indicated otherwise.

T. Herbig, F. Gerl, and W. Minker: Self-Learning Speaker Identification, SCT, pp. 5–57.
springerlink.com

generates the excitation [Campbell, 1997]. The opening of the glottis determines the type of excitation and whether voiced or unvoiced speech is produced. Unvoiced speech is caused by turbulences in the vocal tract whereas voiced speech is due to a quasi-periodic excitation [Schukat-Talamazzini, 1995]. The periodicity of the excitation is called fundamental frequency F0 or pitch. It depends on speaker and gender specific properties of the vocal cords, e.g. length, tension and mass [Campbell, 1997]. The fundamental frequencies of male speakers fall in the range between 80 Hz and 160 Hz and are on average about 132 Hz. Female speakers have an average fundamental frequency of 223 Hz. Children use even higher frequencies.

The *Vocal Tract* (VT) generally comprises all articulators above the vocal cords that are involved in the speech production process [Campbell, 1997]. It can be roughly approximated by a series of acoustic tubes producing characteristic resonances known as formant frequencies [Campbell, 1997; Rabiner and Juang, 1993]. As speakers differ anatomically in shape and length of their vocal tract, formant frequencies are speaker dependent [Campbell, 1997]. The length of the vocal tract is given by the distance between the glottis and the lips. Male speakers have an average length of 17 cm whereas the vocal tract of female speakers is 13 cm long on average.

The modeling of speech production is based on the three main components comprising excitation source, vocal tract and lip radiation [Schukat-Talamazzini, 1995]. The characteristics of glottis, vocal tract and lip radiation are modeled by filters with the impulse responses h_{G}, h_{VT} and h_{LR}. In this context $u(\tau)$ denotes the excitation signal at time instant τ. The speech signal $s(\tau)$ is given by the convolution

$$s(\tau) = u(\tau) * h_{\mathrm{G}}(\tau) * h_{\mathrm{VT}}(\tau) * h_{\mathrm{LR}}(\tau) \tag{2.1}$$

$$u(\tau) * h(\tau) = \int_{\tilde{\tau}=-\infty}^{\infty} u(\tilde{\tau}) \cdot h(\tau - \tilde{\tau}) \, \mathrm{d}\tilde{\tau}. \tag{2.2}$$

Fig. 2.1 depicts the associated block diagram of the source-filter model. A simplified model can be given by

$$s(\tau) = u(\tau) * h_{\mathrm{VT}}(\tau) \tag{2.3}$$

if the glottis and the lip radiation are neglected [Wendemuth, 2004].

Additive noise and channel characteristics, e.g. the speech transmission from the speaker's mouth to the microphone, can be integrated by $n(\tau)$ and a further room impulse response h_{Ch}. The speech signal s_{Mic} recorded by the microphone is given by

$$s_{\mathrm{Mic}}(\tau) = u(\tau) * \underbrace{h_{\mathrm{VT}}(\tau) * h_{\mathrm{Ch}}(\tau)}_{h(\tau)} + n(\tau). \tag{2.4}$$

Speaker variability can be viewed as a result of gender and speaker dependent excitation, anatomical differences in the vocal tract and acquired speaking habits [Campbell, 1997]. Low-level acoustic information is related to the vocal apparatus whereas higher-level information is attributed to learned habits and style, e.g. prosody, word usage and conversational style [Reynolds et al., 2003].

Noisy environments can have an additional influence on the manner of articulation which is known as Lombard effect [Rabiner and Juang, 1993]. Since speakers try to improve communication intelligibility, significant changes in their voice patterns can occur degrading automated speech recognition and speaker identification [Goldenberg et al., 2006]. The effects are highly speaker-dependent and cause an increase in volume, fundamental frequency and vowel duration as well as a shift of the first two formants and energy distribution [Junqua, 1996].

Speaker change detection and speaker identification have to provide appropriate statistical models and techniques which optimally capture the individual speaker characteristics.

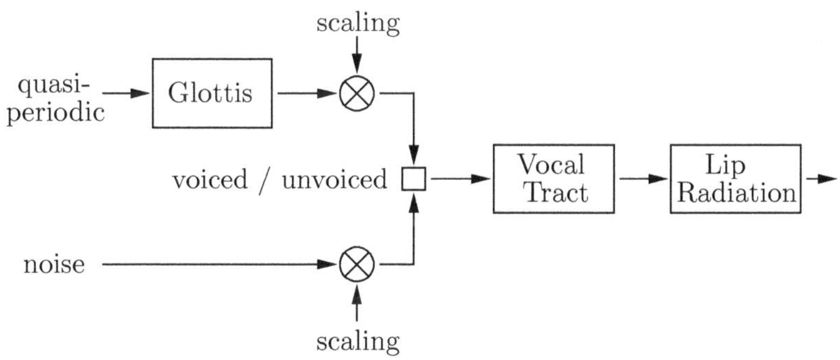

Fig. 2.1 Source-filter model for voiced and unvoiced speech production as found by Rabiner and Juang [1993]; Schukat-Talamazzini [1995]. The speech production comprises either a quasi-periodic or noise-like excitation and three filters which represent the characteristics of the glottis, vocal tract and the lip radiation.

The focus of speech recognition is to understand the content of the spoken phrase. Therefore further aspects of speech are important. Speech can be decomposed into phonemes[2] which can be discriminated by their place of articulation as exemplified for vowels, fricatives, nasal consonants and plosives:

- Vowels are produced by a quasi-periodic excitation at the vocal cords and are voiced. They are characterized by line spectra located at multiples of the fundamental frequency and their intensity exceeds other phonemes.

[2] Phonemes denote the smallest linguistic units of a language. Many languages can be described by 20 − 40 phonemes. The physical sound generated by the articulation of a phoneme is called phone [O'Shaughnessy, 2000].

- Nasal consonants are characterized by a total constriction of the vocal tract and a glottal excitation [Rabiner and Juang, 1993]. The nasal cavity is excited and attenuates the sound signals so that nasals are less intensive than vowels.
- Fricatives are produced by constrictions in the vocal tract that cause a turbulent noisy airflow. The constriction produces a loss of energy so that fricatives are less intensive. They are characterized by energy located at higher frequencies.
- Plosives (stops) are produced by an explosive release of an occlusion of the vocal tract followed by a turbulent noisy air-stream.

The phoneme based speaker identification discussed in Sect. 3.3 exploits the knowledge about the speaker variability of phonemes to enhance the identification accuracy. For example, no information about the vocal tract is present if the place of articulation is given by the teeth or lips as in /f/ or /p/. In consequence, little speaker discrimination is expected.

In the next section an automated method for feature extraction is presented. The goal is to extract the relevant spectral properties of speech which can be used for speaker change detection, speaker identification and speech recognition.

2.2 Front-End

The front-end is the first step in the speech recognition or speaker identification processing chain. In this book it consists of a signal processing part, feature extraction and post-processing as displayed in Fig. 2.2.

The signal processing applies a sampling and preprocessing to the recorded microphone signals. The feature extraction transforms discrete time signals into a vector representation which is more feasible for pattern recognition algorithms. The following post-processing comprises feature vector normalization to compensate for channel characteristics. In the case of speech recognition a discriminative mapping is used in addition. The three parts of the front-end are described in more detail.

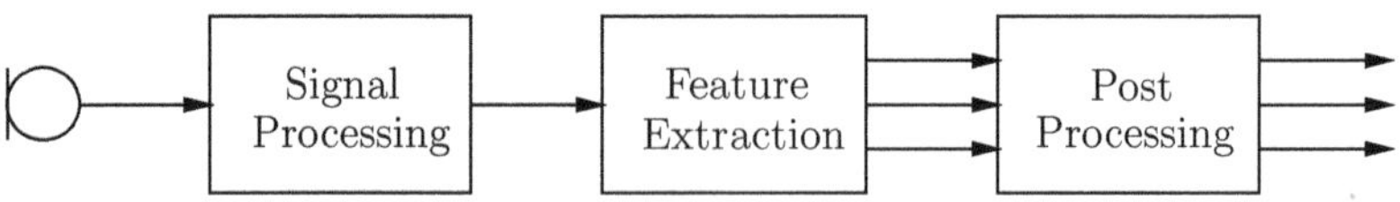

Fig. 2.2 Block diagram of a front-end for speech recognition and speaker identification. The recorded microphone signal is preprocessed to reduce background noises. A vector representation is extracted from the speech signal. The feature vectors are normalized in the post processing to compensate for channel characteristics.

Signal Processing

The signal processing receives an input signal from the microphone and delivers a digitized and enhanced speech signal to feature extraction. The principle block diagram is displayed in Fig. 2.3.

Fig. 2.3 Block diagram of the signal processing in the front-end. The recorded speech signal is sampled at discrete time instances and noise reduction is performed. Speech pauses are excluded for subsequent speech processing.

The *Analog to Digital Converter* (ADC) samples the incoming time signals at discrete time instances

$$s_l^{\mathrm{Mic}} = s_{\mathrm{Mic}}(\frac{l}{f_{\mathrm{s}}}), \quad l = 0, 1, 2, 3, \ldots \tag{2.5}$$

and performs a quantization. In this book a sampling rate of $f_{\mathrm{s}} = 11.025\,\mathrm{kHz}$ and a 16 bit quantization are used. In this context l denotes the discrete time index.

Especially in automotive applications, speech signals are superimposed with background noises affecting the discrimination of different speakers or speech decoding. Noise reduction targets to minimize environmental influences which is essential for a reliable recognition accuracy.

Noise reduction can be achieved by a spectral decomposition and a spectral weighting as described by Vary et al. [1998]. The time signal is split into its spectral components by an analysis filter bank. A noise estimation algorithm calculates the noise spectrum as found by Cohen [2003]; Cohen and Berdugo [2002]. In combination with an estimate of the disturbed speech spectrum, the power of the undisturbed or clean speech signal is estimated by a spectral weighting. The Wiener filter, spectral subtraction and Ephraim-Malah are well-known weighting techniques which are described by Cappé [1994]; Ephraim and Malah [1984] in more detail. The phase of the speech signals remains unchanged since phase distortions seem to be less critical for human hearing [Vary et al., 1998]. After spectral weighting the temporal speech signals are synthesized by a synthesis filter bank. The enhanced speech signal is employed in the subsequent feature extraction. Fig. 2.4 displays the noise reduction setup as described above.

As noise reduction algorithms do not play an important role in this book, only the widely-used Wiener filter is applied. Interested readers are referred to Benesty et al. [2005]; Hänsler and Schmidt [2004, 2008] for further detailed information.

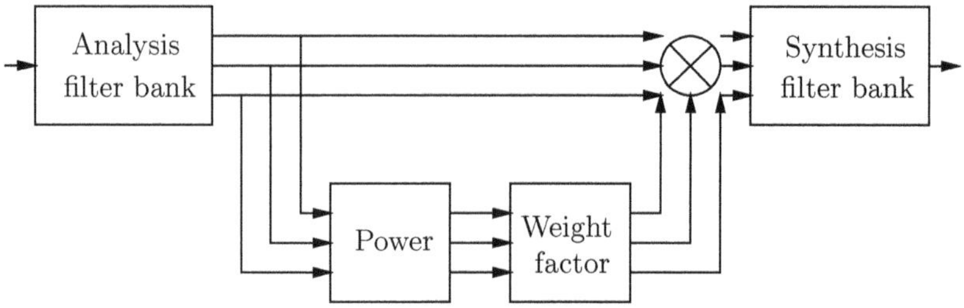

Fig. 2.4 Block diagram of the noise reduction. The signal is split into its spectral components by an analysis filter bank. Noise and speech power are estimated to calculate a spectral weighting. The power of the clean speech signal is estimated by a spectral weighting of the disturbed speech spectrum. The enhanced speech signal is synthesized by the synthesis filter bank.

Voice Activity Detection (VAD) or speech segmentation excludes speech pauses so that the subsequent processing is done on portions of the microphone signal that are assumed to comprise only speech utterances. The computational load and the probability of false classifications can be reduced for speaker identification and speech recognition. Conventional speech segmentation algorithms rely on speech properties such as the zero-crossing rate, rising or falling energy[3] $\mathrm{E}\{(s_l^{\mathrm{Mic}})^2\}$ [Kwon and Narayanan, 2005]. Furthermore, the harmonic structure of voiced speech segments may be used, especially for vowels. Adverse environments complicate the end-pointing because background noises mask regions of low speech activity. Finally, speech segmentation can shift the detected boundaries of the beginning and end of an utterance to guarantee that the entire phrase is enclosed for speech recognition. For further reading, the interested reader is referred to literature such as Espi et al. [2010]; Ramírez et al. [2007].

Feature Extraction

Feature extraction transforms the enhanced speech signals into a vector representation which reflects discriminatory properties of speech. The *Mel Frequency Cepstral Coefficients* (MFCC) are frequently used in speech recognition and speaker identification [O'Shaughnessy, 2000; Quatieri, 2002]. MFCCs are physiologically motivated by human hearing[4].

The reader may wonder why the MFCC features are also used for speaker identification. Even though MFCCs are expected to cause a loss of speaker information, they seem to capture enough spectral information for speaker identification. The vocal tract structure may be considered as the

[3] E{} denotes the expectation value.

[4] For the comparison of human and machine in speaker identification and speech recognition it is referred to Hautamäki et al. [2010]; Liu et al. [1997]; O'Shaughnessy [2000] and Meyer et al. [2006].

dominant physiological feature to distinguish speakers. This feature is reflected by the speech spectrum [Reynolds and Rose, 1995]. Kinnunen [2003] provides a thorough investigation of several spectral features for speaker identification and concludes that the best candidates range at the same level. However, features directly computed by a filter bank such as the MFCCs are obviously better suited for noisy applications [Reynolds and Rose, 1995]. MFCCs enable a compact representation and are suitable for statistical modeling by Gaussian mixtures [Campbell, 1997] as explained later in Sect. 2.4. Reynolds [1995a,b] obtains excellent speaker identification rates even for large speaker populations. According to Quatieri [2002] the mel-cepstrum can be regarded as one of the most effective features for speech-related pattern recognition.

Hence, the description in this book is restricted to the extraction of the MFCC features. High-level features [Quatieri, 2002; Reynolds et al., 2003] are not considered since a command and control scenario is investigated. An exemplary MFCC block diagram is shown in Fig. 2.6.

First, the *Short Time Fourier Transform* (STFT) divides the sequence of speech samples in frames of predetermined length, applies a window function and splits each frame into its spectral components. Common window functions are the Hamming and Hann window [Vary et al., 1998]. In this book the window function $\tilde{h}$ is realized by the Hann window

$$\tilde{h}_l = \frac{1}{2}\left(1 + \cos(\frac{2\pi l}{N_{\text{Frame}}})\right), \qquad l = -\frac{1}{2}N_{\text{Frame}}, \ldots, \frac{1}{2}N_{\text{Frame}}, \tag{2.6}$$

where the length of one frame is given by N_{Frame}. Subsequently, t denotes the discrete frame index. The frame shift N_{shift} describes the time which elapsed between two frames. The frame length is 20 ms and frame shift is given as half of the frame length. The windowing

$$\tilde{s}^{\text{w}}_{t,l} = s^{\text{Mic}}_{t \cdot N_{\text{shift}}+l} \cdot \tilde{h}_l, \qquad t > 1, \tag{2.7}$$

combined with the *Discrete Fourier Transform* (DFT) or *Fast Fourier Transform* (FFT)[5] realizes a filter bank [Vary et al., 1998]. The filter characteristics are determined by the window's transfer function shifted in the frequency domain [Hänsler and Schmidt, 2004; Vary et al., 1998]. The incoming microphone signals are split into N_{FFT} narrow band signals $S(f_{\text{b}}, t)$ [Hänsler and Schmidt, 2004; Rabiner and Juang, 1993]:

$$\begin{aligned} S(f_{\text{b}}, t) &= \mathcal{FFT}\left\{\tilde{s}^{\text{w}}_{t,l}\right\} \\ &= \sum_{l=-\frac{1}{2}N_{\text{Frame}}}^{\frac{1}{2}N_{\text{Frame}}} \tilde{s}^{\text{w}}_{t,l} \cdot \exp(-\mathrm{i}\frac{2\pi}{N_{\text{FFT}}} f_{\text{b}}\, l), \quad N_{\text{FFT}} \geq N_{\text{Frame}}. \end{aligned} \tag{2.8}$$

[5] The FFT represents an efficient implementation of the DFT [Kammeyer and Kroschel, 1998]. Details on DFT and FFT are given by Kammeyer and Kroschel [1998]; Oppenheim and Schafer [1975], for example.

In this context the index f_b denotes the frequency bin and i is the imaginary unit. The FFT order is selected as $N_{\mathrm{FFT}} = 256$ and zero-padding [Oppenheim and Schafer, 1975] is applied.

Applying the STFT to speech signals presumes that speech can be regarded as stationary for short time periods. Stationarity claims that the statistical properties of the random variable are independent from the investigated time instances [Hänsler, 2001]. Irrespective of voiced or unvoiced speech, the spectral amplitudes of speech can be regarded as quasi-stationary for tens of milliseconds [O'Shaughnessy, 2000]. For the feature extraction this assumption is approximately true if the frame length is shorter than 30 ms [Schukat-Talamazzini, 1995].

In the next step of the feature extraction the magnitude $|S(f_b, t)|$ and the local energy $|S(f_b, t)|^2$ are calculated. The phase information is discarded.

The human auditory system cannot resolve frequencies linearly [Logan, 2000; Wendemuth, 2004]. The critical-band rate or Bark scale reflects the relationship between the actual and frequency resolution which is approximated by the mel filter bank [O'Shaughnessy, 2000]. The mel filter bank comprises a non-linear frequency warping and a dimension reduction. The relation between linear frequencies f and warped mel-frequencies f_{mel} is derived from psychoacoustic experiments and can be mathematically described by

$$f_{\mathrm{mel}} = 2595 \cdot \log_{10}(1 + \frac{f}{700\,\mathrm{Hz}}) \tag{2.9}$$

as found by O'Shaughnessy [2000]. Up to 1 kHz the linear and warped frequencies are approximately equal whereas higher frequencies are logarithmically compressed.

In the warped frequency domain an average energy is computed by an equally spaced filter bank [Mammone et al., 1996]. In the linear frequency domain this corresponds to filters with increasing bandwidth for higher frequencies. Triangular filters can be employed as found by Davis and Mermelstein [1980]. The spectral resolution decreases with increasing frequency. In Fig. 2.5 an example is shown. Further details can be found by Rabiner and Juang [1993].

The number of frequency bins is reduced here by this filter bank from $\frac{1}{2}N_{\mathrm{FFT}} + 1$ to 19. The result can be grouped into the vectors $\mathbf{x}_t^{\mathrm{MEL}}$ representing all energy values of frame index t. For convenience, all vectors are viewed as column vectors.

The logarithm is applied element by element to $\mathbf{x}_t^{\mathrm{MEL}}$ and returns $\mathbf{x}_t^{\mathrm{LOG}}$. The dynamics of the mel energies are compressed and products of spectra are deconvolved [O'Shaughnessy, 2000]. For example, speech characteristics and scaling factors are separated into additive components in the case of clean speech.

The *Discrete Cosine Transform* (DCT) is a well-known algorithm from image processing because of its low computational complexity and decorrelation property. The DCT approximates the Karhuen-Loeve Transform (KLT)

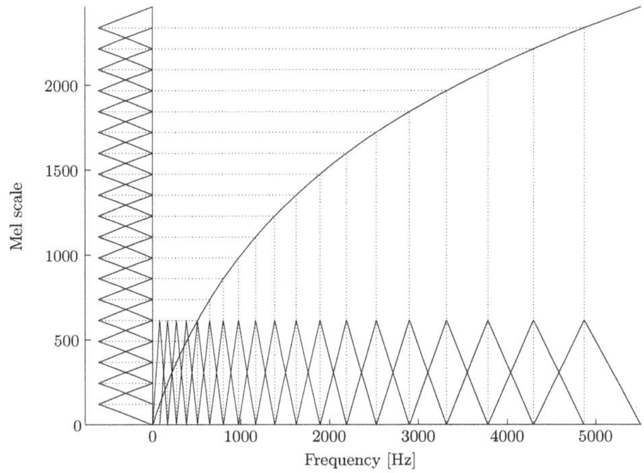

Fig. 2.5 Example for a mel filter bank. An equally spaced filter bank is used in the warped frequency domain to compute an averaged energy. In the linear frequency domain the bandwidth of the corresponding filters increases for higher frequencies. Triangular filters can be applied, for example.

in the case of Markov random signals of first order [Britanak et al., 2006] which are described later in Sect. 2.5.2. The KLT also known as *Principal Component Analysis* (PCA) projects input vectors onto directions efficient for representation [Duda et al., 2001]. The DCT can be combined with a dimension reduction by omitting the higher dimensions.

The DCT in (2.10) is applied to the logarithmic mel-energies. The coefficients $x_{l,t}^{\mathrm{DCT}}$ of the resulting vector $\mathbf{x}_t^{\mathrm{DCT}}$ are given by the formula

$$x_{l,t}^{\mathrm{DCT}} = \sum_{l_1=1}^{d_{\mathrm{LOG}}} x_{l,t}^{\mathrm{LOG}} \cdot \cos\left(l(l_1 - \frac{1}{2})\frac{\pi}{d_{\mathrm{LOG}}} \right), \quad l = 1, \ldots, d_{\mathrm{DCT}}, \tag{2.10}$$

as found by O'Shaughnessy [2000]. d contains the dimensionality of the specified vector. The first coefficient ($l = 0$) is typically not used because it represents an average energy of $\mathbf{x}_t^{\mathrm{LOG}}$ and therefore depends on the scaling of the input signal [O'Shaughnessy, 2000]. It is replaced in this book by a logarithmic energy value which is derived from $\mathbf{x}_t^{\mathrm{MEL}}$. The second coefficient of the DCT can be interpreted as the ratio of the energy located at low and high frequencies [O'Shaughnessy, 2000]. The resulting vector is denoted by $\mathbf{x}_t^{\mathrm{MFCC}}$. A dimension reduction is used for all experiments in this book and the remaining 11 MFCC features are employed.

Post Processing

In post processing a normalization of the developed feature vectors $\mathbf{x}_t^{\mathrm{MFCC}}$ is performed and a *Linear Discriminant Analysis* (LDA) is applied. Fig. 2.7

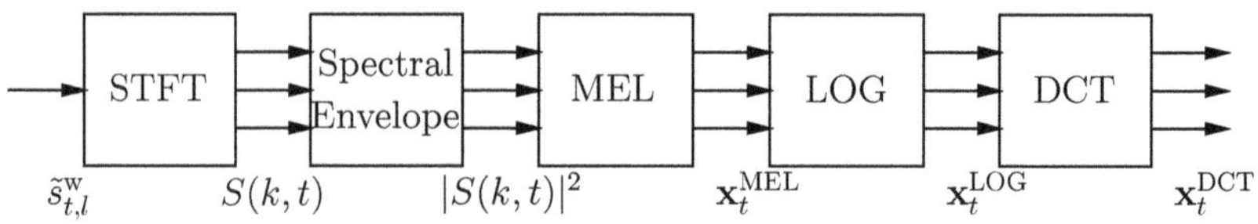

Fig. 2.6 Block diagram of the MFCC feature extraction. The STFT is applied to the windowed speech signal and the spectral envelope is further processed by the mel filter bank. The discrete cosine transform is applied to the logarithm of the mel energies. The MFCC feature vectors $\mathbf{x}_t^{\mathrm{MFCC}}$ are usually computed by discarding the first coefficient of $\mathbf{x}_t^{\mathrm{DCT}}$. A dimension reduction can be applied.

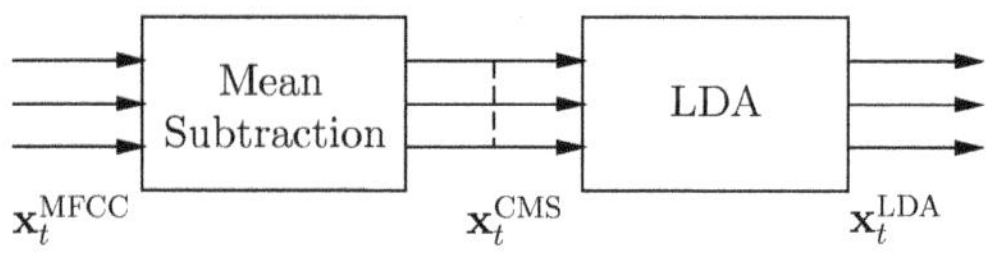

Fig. 2.7 Block diagram of the post processing in the front-end. The cepstral mean vector iteratively estimated is subtracted from the MFCC vectors. In the case of a speech recognizer an LDA is employed to increase the discrimination of clusters in feature space, e.g. phonemes.

displays the corresponding block diagram comprising the *Cepstral Mean Subtraction* (CMS) and LDA.

Mean subtraction is an elementary approach to compensate for channel characteristics [Mammone et al., 1996]. It improves the robustness of speaker identification and speech recognition [Reynolds et al., 2000; Young et al., 2006]. For common speech recognizers it has been established as a standard normalization method [Buera et al., 2007].

Noise reduction as a pre-processing diminishes the influence of background noises on the speech signal whereas the room impulse response h_{Ch} in (2.4) remains relatively unaffected. The logarithm of feature extraction splits the speech signal and the room impulse response into additive components if the channel transfer function is spectrally flat within the mel filters and if clean speech is considered. When the room impulse response varies only slowly in time relative to the speech signal, the long-term average of feature vector $\boldsymbol{\mu}^{\mathrm{MFCC}}$ is approximately based on the unwanted channel characteristics. The subtraction of this mean vector from the feature vectors compensates for unwanted channel characteristics such as the microphone transfer function and leaves the envelope of the speech signal relatively unaffected. In addition, the inter-speaker variability is reduced by mean normalization [Häb-Umbach, 1999].

To avoid latencies caused by the re-processing of entire utterances the mean can be continuously adapted with each recorded utterance to track the long-term average. A simple iterative mean computation is given by

$$\boldsymbol{\mu}_t^{\mathrm{MFCC}} = \frac{n_{\mathrm{MFCC}} - 1}{n_{\mathrm{MFCC}}} \boldsymbol{\mu}_{t-1}^{\mathrm{MFCC}} + \frac{1}{n_{\mathrm{MFCC}}} \mathbf{x}_t^{\mathrm{MFCC}}, \quad t > 1 \tag{2.11}$$

$$\boldsymbol{\mu}_1^{\mathrm{MFCC}} = \mathbf{x}_1^{\mathrm{MFCC}} \tag{2.12}$$

where the total number of feature vectors is denoted by n_{MFCC}.

Class et al. [1993, 1994] describe an advanced method to iteratively adapt the mean subtraction. The mean of $\mathbf{x}_t^{\mathrm{MFCC}}$ is realized by a recursion of first order. A decay factor is chosen so that a faster adaptation during the enrollment of new speakers is achieved. A slower constant learning rate is guaranteed afterwards.

In this book, the mean vector $\boldsymbol{\mu}_t^{\mathrm{MFCC}}$ is estimated according to (2.11). $\boldsymbol{\mu}_1^{\mathrm{MFCC}}$ and n_{MFCC} are initialized in an off-line training. Furthermore, n_{MFCC} is limited to guarantee a constant learning rate in the long run. Mean normalization is not applied to the first coefficient ($l = 0$) which is replaced by a logarithmic energy estimate as explained above. A peak tracker detecting the maxima of the energy estimate is used for normalization.

Further normalization techniques can be found by Barras and Gauvain [2003]; Segura et al. [2004].

MFCCs capture only static speech features since the variations of one frame are considered [Young et al., 2006]. The dynamic behavior of speech can be partially captured by the first and second order time derivatives of the feature vectors. These features are also referred as delta features $\mathbf{x}_t^{\Delta}$ and delta-delta features $\mathbf{x}_t^{\Delta^2}$ and are known to significantly enhance speech recognition accuracy [Young et al., 2006]. Delta features are computed by the weighted sum of feature vector differences within a time window of $2 \cdot N_\Delta$ frames given by

$$\mathbf{x}_t^{\Delta} = \frac{\sum_{l=1}^{N_\Delta} l \cdot \left(\mathbf{x}_{t+l}^{\mathrm{MFCC}} - \mathbf{x}_{t-l}^{\mathrm{MFCC}}\right)}{2 \cdot \sum_{j=l}^{N_\Delta} l^2} \tag{2.13}$$

as found by Young et al. [2006]. The same algorithm can be applied to the delta features to compute delta-delta features. The feature vectors used for speech recognition may comprise static and dynamic features stacked in a supervector.

Delta or delta-delta features are not computed in the speech recognizer used for the experiments of this book. Instead, the preceding and subsequent 4 vectors are stacked together with the actual feature vector in a long supervector. The dynamics of delta and delta-delta coefficients have been incorporated into the LDA using a bootstrap training [Class et al., 1993].

An LDA is applied in order to obtain a better discrimination of clusters in feature space, e.g. phonemes. It is a linear transform realized by a matrix multiplication. Input vectors are projected onto those directions which are efficient for discrimination [Duda et al., 2001]. The result is a compact representation of each cluster with an improved spatial discrimination with respect to other clusters. Furthermore, a dimension reduction can be achieved.

Further details can be found by Duda et al. [2001]. The speech decoding works on the resulting vectors $\mathbf{x}_t^{\mathrm{LDA}}$ as explained later in Sect. 2.5.

An unsupervised system for speaker identification usually encounters difficulties to define reasonable clusters. The most obvious strategy is to cluster the feature vectors of each speaker separately. Since the pool of speakers is generally unknown a priori, speaker identification usually does not use an LDA.

2.3 Speaker Change Detection

The feature extraction described in the preceding section provides a compact representation of speech that allows further automated processing.

One application is speaker change detection which performs a segmentation of an incoming speech signal into homogeneous parts in an unsupervised way. The goal is to obtain precise boundaries of the recorded utterances so that always only one speaker is enclosed. Then speaker identification and speech recognition can be applied to utterances containing only one speaker.

In Sect. 3.1 several strategies for audio signal segmentation are outlined. Many algorithms work on the result of a speaker change detection as the first part of an unsupervised speaker tracking.

First, the motivation of speaker change detection is introduced and the corresponding problem is mathematically described. Then a detailed description of one representative which is well-known from literature is presented.

2.3.1 Motivation

The task of speaker change detection is to investigate a stream of audio data for possible speaker turns. For example, a buffered sequence of a data stream is divided into segments of fixed length and each boundary has to be examined for speaker turns. Thus, a representation of speech data suitable for speaker change detection has to be found and a strategy has to be developed to confirm or reject hypothetical speaker changes.

The front-end introduced in Sect. 2.2 extracts the relevant features from the time signal and provides a continuous stream of MFCC feature vectors for speech recognition and speaker identification. Ajmera et al. [2004] demonstrate that MFCCs can also be used to detect speaker changes. Thus, only MFCCs are considered to limit the computational complexity of the target system including self-learning speaker identification and speech recognition. However, other features could be used as well in the subsequent considerations.

In the following, a data stream is iteratively investigated frame by frame or in data blocks for speaker turns. In Fig. 2.8 an example is shown. A sequence of T frames is given in a buffer. At time instance t_{Ch} a speaker change is hypothesized and the algorithm has to decide whether part I and part II

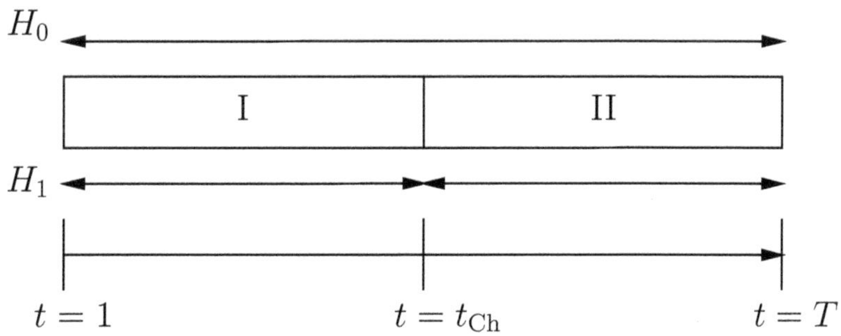

Fig. 2.8 Problem of speaker change detection. A continuous data stream has to be examined for all candidates where a speaker turn can occur. The basic procedure checks a given time interval by iteratively applying metric-based, model-based or decoder-guided techniques.

originate from one or two speakers. This procedure can be repeated for all possible time instances $1 < t_{\mathrm{Ch}} < T$. In Sect. 3.1 several strategies are listed to select hypothetical speaker changes.

Here it is assumed that a hypothetical speaker turn at an arbitrary time instance t_{Ch} has to be examined and that a binary hypothesis test has to be done. Subsequently, the shorthand notation $\mathbf{x}_{1:t_{\mathrm{Ch}}}$ for the sequence of feature vectors $\{\mathbf{x}_1, \ldots, \mathbf{x}_{t_{\mathrm{Ch}}}\}$ is used.

- **Hypothesis** H_0 assumes no speaker change so that both sequences of feature vectors $\mathbf{x}_{1:t_{\mathrm{Ch}}}$ and $\mathbf{x}_{t_{\mathrm{Ch}}+1:T}$ originate from one speaker.
- **Hypothesis** H_1 presumes a speaker turn so that two speakers are responsible for $\mathbf{x}_{1:t_{\mathrm{Ch}}}$ and $\mathbf{x}_{t_{\mathrm{Ch}}+1:T}$.

The optimal decision of binary hypothesis tests can be realized by the Bayesian framework. In the case of two competing hypotheses H_0 and H_1, H_0 is accepted by the Bayesian criterion if the following inequality is valid [Hänsler, 2001]:

$$\frac{p(\mathbf{x}_{1:T}|H_0)}{p(\mathbf{x}_{1:T}|H_1)} \overset{H_0}{\geq} \frac{(c_{10} - c_{11}) \cdot p(H_1)}{(c_{01} - c_{00}) \cdot p(H_0)}. \tag{2.14}$$

Here the parameters c_{01}, c_{00}, c_{10} and c_{11} denote the costs for falsely accepted H_1, truly detected H_0, falsely accepted H_0 and correctly detected H_1, respectively. The probabilities $p(H_0)$ and $p(H_1)$ stand for the prior probabilities of the specified hypothesis.

The left side of (2.14) is called the likelihood ratio. The right side contains the prior knowledge and acts as a threshold. If the costs for correct decisions equal zero and those for false decisions are equal, the threshold only depends on the prior probabilities. For equal prior probabilities the criterion chooses the hypothesis with the highest likelihood.

A statistical model is required to compute the likelihoods $p(\mathbf{x}_{1:T}|H_0)$ and $p(\mathbf{x}_{1:T}|H_1)$. The existing algorithms for finding speaker turns can be classified into metric-based, model-based and decoder-guided approaches as found by Chen and Gopalakrishnan [1998].

In this context only one prominent representative of the model-based algorithms is introduced and discussed. For further details, especially to metric-based or decoder-guided algorithms, the interested reader is referred to the literature listed in Sect. 3.1.

2.3.2 Bayesian Information Criterion

The so-called *Bayesian Information Criterion* (BIC) belongs to the most commonly applied techniques for speaker change detection [Ajmera et al., 2004]. There exist several BIC variants and therefore only the key issues are discussed in this section.

Fig. 2.8 denotes the first part before the assumed speaker turn as I and the second part as II. The binary hypothesis test requires the statistical models for part I and II as well as for the conjunction of both parts I + II. The statistical models for part I and II cover the case of two origins, e.g. two speakers, and thus reflect the assumption of hypothesis H_1. The model built on the conjunction presumes only one origin as it is the case in H_0.

Parameter Estimation

Each part is modeled by one multivariate Gaussian distribution which can be robustly estimated even on few data. Subsequently, $\boldsymbol{\mu}$, Σ, $|\Sigma|$ and d denote the mean vector, covariance matrix, the determinant of the covariance matrix and the dimension of the feature vectors, respectively. The superscript indices T and -1 are the transpose and inverse of a matrix. Θ is the parameter set of a multivariate Gaussian distribution comprising the mean vector and covariance. The probability density function is given by

$$\begin{aligned} p(\mathbf{x}_t|\Theta) &= \mathcal{N}\{\mathbf{x}_t|\boldsymbol{\mu},\Sigma\} \\ &= \frac{1}{(2\pi)^{0.5d}\,|\Sigma|^{0.5}} \cdot \mathrm{e}^{-\frac{1}{2}(\mathbf{x}_t-\boldsymbol{\mu})^T\cdot\Sigma^{-1}\cdot(\mathbf{x}_t-\boldsymbol{\mu})}. \end{aligned} \tag{2.15}$$

In addition, all features are usually assumed to be *independent and identically distributed* (iid) random variables. Even though successive frames of the speech signal are not statistically independent as shown in Sect. 2.5, this assumption simplifies parameter estimation and likelihood computation but is still efficient enough to capture speaker changes. Since the statistical dependencies of the speech trajectory are neglected, the basic statistical models can be easily estimated for each part of the data stream.

The mean vector and covariance matrix are unknown a priori and can be determined by the *Maximum Likelihood* (ML) estimates [Bishop, 2007; Lehn and Wegmann, 2000]. An ergodic random process is assumed so that the expectation value can be replaced by the time average [Hänsler, 2001]. The parameters for the statistical model of the hypothesis H_0 are given by

$$\boldsymbol{\mu}_{\mathrm{I+II}} = \mathrm{E}_{\mathbf{x}}\{\mathbf{x}_t\} \approx \frac{1}{T}\sum_{t=1}^{T}\mathbf{x}_t \tag{2.16}$$

$$\Sigma_{\mathrm{I+II}} = \mathrm{E}_{\mathbf{x}}\{(\mathbf{x}_t - \boldsymbol{\mu}_{\mathrm{I+II}}) \cdot (\mathbf{x}_t - \boldsymbol{\mu}_{\mathrm{I+II}})^T\} \tag{2.17}$$

$$\approx \frac{1}{T}\sum_{t=1}^{T}(\mathbf{x}_t - \boldsymbol{\mu}_{\mathrm{I+II}}) \cdot (\mathbf{x}_t - \boldsymbol{\mu}_{\mathrm{I+II}})^T. \tag{2.18}$$

In the case of a speaker turn, the parameters are estimated separately on part I and II as given by

$$\boldsymbol{\mu}_{\mathrm{I}} = \mathrm{E}_{\mathbf{x}}\{\mathbf{x}_t\} \approx \frac{1}{t_{\mathrm{Ch}}}\sum_{t=1}^{t_{\mathrm{Ch}}}\mathbf{x}_t \tag{2.19}$$

$$\Sigma_{\mathrm{I}} = \mathrm{E}_{\mathbf{x}}\{(\mathbf{x}_t - \boldsymbol{\mu}_{\mathrm{I}}) \cdot (\mathbf{x}_t - \boldsymbol{\mu}_{\mathrm{I}})^T\} \tag{2.20}$$

$$\approx \frac{1}{t_{\mathrm{Ch}}}\sum_{t=1}^{t_{\mathrm{Ch}}}(\mathbf{x}_t - \boldsymbol{\mu}_{\mathrm{I}}) \cdot (\mathbf{x}_t - \boldsymbol{\mu}_{\mathrm{I}})^T. \tag{2.21}$$

The parameters $\boldsymbol{\mu}_{\mathrm{II}}$ and Σ_{II} are estimated on part II in analogy to $\boldsymbol{\mu}_{\mathrm{I}}$ and Σ_{I}.

The likelihood computation of feature vector sequences becomes feasible due to the iid assumption. The log-likelihood can be realized by the sum of log-likelihoods of each time step

$$\log\left(p(\mathbf{x}_{1:T}|\Theta)\right) = \sum_{t=1}^{T}\log\left(p(\mathbf{x}_t|\Theta)\right). \tag{2.22}$$

Now the likelihood values $p(\mathbf{x}_{1:T}|H_0)$ and $p(\mathbf{x}_{1:T}|H_1)$ can be given for both hypotheses. For convenience, only log-likelihoods are examined

$$\log\left(p(\mathbf{x}_{1:T}|H_0)\right) = \sum_{t=1}^{T}\log(\mathcal{N}\{\mathbf{x}_t|\boldsymbol{\mu}_{\mathrm{I+II}}, \Sigma_{\mathrm{I+II}}\}) \tag{2.23}$$

$$\log\left(p(\mathbf{x}_{1:T}|H_1)\right) = \sum_{t=1}^{t_{\mathrm{Ch}}}\log(\mathcal{N}\{\mathbf{x}_t|\boldsymbol{\mu}_{\mathrm{I}}, \Sigma_{\mathrm{I}}\}) + \sum_{t=t_{\mathrm{Ch}}+1}^{T}\log(\mathcal{N}\{\mathbf{x}_t|\boldsymbol{\mu}_{\mathrm{II}}, \Sigma_{\mathrm{II}}\}), \tag{2.24}$$

whereas the parameter set Θ is omitted.

Hypothesis Test

The hypothesis test can now be realized by using these likelihoods and a threshold

$$\theta_{\text{th}} \lessgtr \log\left(p(\mathbf{x}_{1:T}|H_1)\right) - \log\left(p(\mathbf{x}_{1:T}|H_0)\right). \quad (2.25)$$

θ_{th} replaces the costs and prior probabilities in (2.14) and has to be determined experimentally. This approach is known as the *Log Likelihood Ratio* (LLR) algorithm [Ajmera et al., 2004].

The BIC criterion test extends this technique by introducing a penalty term and does not need a threshold θ_{th}. The hypothesis H_1 assumes a speaker turn and therefore applies two Gaussians in contrast to H_0. The models of H_1 are expected to lead to a better representation of the observed speech data since the number of model parameters is doubled. The penalty term P compensates this inequality and guarantees an unbiased test [Ajmera et al., 2004]. A speaker turn is assumed if the following inequality is valid

$$\log\left(p(\mathbf{x}_{1:T}|H_1)\right) - \log\left(p(\mathbf{x}_{1:T}|H_0)\right) - \mathrm{P} \geq 0, \quad (2.26)$$

where the penalty

$$P = \frac{1}{2} \cdot \left(d + \frac{1}{2}d(d+1)\right) \cdot \log T \quad (2.27)$$

comprises the number of parameters which are used to estimate the mean and covariance. P may be multiplied by a tuning parameter λ which should theoretically be $\lambda = 1$ as found by Ajmera et al. [2004].

The training and test data of the Gaussian distributions are identical and permit a more efficient likelihood computation. The BIC criterion only requires to compute the covariances for both parts and their conjunction as found by Zhu et al. [2005]. Thus, the inequality

$$T \cdot |\Sigma_{\mathrm{I+II}}| - t_{\mathrm{Ch}} \cdot |\Sigma_{\mathrm{I}}| - (T - t_{\mathrm{Ch}}) \cdot |\Sigma_{\mathrm{II}}| - \lambda \cdot P \leq 0 \quad (2.28)$$

is also known as the variance BIC [Nishida and Kawahara, 2005]. If the factor λ is equal to zero, equation (2.28) describes the LLR algorithm [Ajmera et al., 2004].

The requirement to estimate robust parameters on limited data does not allow using more complex statistical models. Even though Gaussian distributions only consist of one mean vector and one covariance matrix, short utterances, e.g. $\leq 2\,\text{sec}$, seem not to provide enough information for a reliable estimation as the results of Zhou and Hansen [2000] suggest.

The complexity of the statistical models can be further reduced by using diagonal covariances. The elements of diagonal covariances are equal to zero except the main diagonal. $\Sigma_{\text{diag}}(l_r, l_c)$ denotes the matrix element of row l_r and column l_c:

$$\Sigma^{\text{diag}}_{l_r,l_c} = \delta_{\text{K}}(l_r,l_c) \cdot \sigma_{l_r}, \qquad \delta_{\text{K}}(l_r,l_c) = \begin{cases} 0 \text{ if } l_r \neq l_c \\ 1 \text{ if } l_r = l_c \end{cases} . \tag{2.29}$$

σ and δ_{K} are the scalar variance and Kronecker delta. This assumption is a trade-off between the reduced number of parameters and statistical modeling accuracy.

Zhou and Hansen [2000] suggest to use the Hotelling measure instead of Gaussian distribution for very short utterances. Their algorithm estimates only the mean vectors $\boldsymbol{\mu}_{\text{I}}$, $\boldsymbol{\mu}_{\text{II}}$ and $\boldsymbol{\mu}_{\text{I+II}}$ separately. The covariance $\boldsymbol{\Sigma}_{\text{I+II}}$ replaces $\boldsymbol{\Sigma}_{\text{I}}$, $\boldsymbol{\Sigma}_{\text{II}}$ and simplifies the covariance estimation.

In this context it should be emphasized that the use case of this book is a command and control application. The average duration can be less than 2 sec which limits the use of complex algorithms for speaker change detection. Since short utterances do not cover all phonemes sufficiently, text-dependency can be a problem. The variation of the utterance duration is expected to be high so that the comparison of such utterances seems to be difficult [Nishida and Kawahara, 2005]. One Gaussian distribution might be not sufficient to capture the variations sufficiently [Nishida and Kawahara, 2005] which is undesirable with respect to the limited training data. Alternatively, more sophisticated statistical models incorporating prior knowledge about speech and speaker characteristics can be used as found by Malegaonkar et al. [2007].

Adverse environments such as in-car applications contain time-variant background noises. Due to the low complexity of the statistical models previously introduced, it seems to be difficult to distinguish properly between the changes of the acoustic environment and speaker changes in an unsupervised manner.

2.4 Speaker Identification

The discussion of speaker changes is now extended to speaker identification and the question who is speaking is addressed in this section. Speaker identification is an essential part of the target system since speakers should be tracked across several speaker changes. Speech recognition can then be enhanced for a particular speaker by individual long-term adaptation as explained in the following sections.

Furthermore, a technique is necessary to detect whether a speaker is known to the system. Only in combination with a robust detection of new users can a completely unsupervised system be achieved which is able to enroll speakers without any additional training.

In this section a short introduction into speaker identification is given by a selection of the relevant literature. The problem of speaker identification is formulated and a brief survey of the main applications and strategies is provided in Sect. 2.4.1. The main representative of the statistical models used for speaker identification is explained in Sect. 2.4.2. The training of the statistical model and speaker identification at run-time are described in more detail. Finally, some techniques are discussed to detect unknown speakers.

2.4.1 Motivation and Overview

Speaker identification has found its way into a manifold of speech controlled applications such as automatic transcription of meetings and conferences, information retrieval systems, security applications, access control and law enforcement [Gish and Schmidt, 1994]. The main task is to identify speakers by their voice characteristics.

Probabilistic approaches determine statistical models for each speaker so that this problem can be solved in a statistical framework. Then speaker identification can be realized by the *Maximum A Posteriori* (MAP) criterion [Gish and Schmidt, 1994; Reynolds and Rose, 1995]. Speaker identification has to find the speaker who maximizes the posterior probability of being the origin of the recorded utterance:

$$i_{\text{MAP}} = \arg\max_i \{p(i|\mathbf{x}_{1:T})\}. \tag{2.30}$$

Variable i and i_{MAP} denote the speaker index and the corresponding MAP estimate.

The current utterance is characterized by a sequence of feature vectors $\mathbf{x}_{1:T}$. For example, MFCC features or mean normalized MFCCs computed by the front-end may be employed for speaker identification [Reynolds, 1995a; Reynolds et al., 2000]. Delta features may be used in addition [Reynolds et al., 2000].

The MAP criterion can be expressed depending on the likelihood $p(\mathbf{x}_{1:T}|i)$ and the prior probability $p(i)$ instead of the posterior probability $p(i|\mathbf{x}_{1:T})$ by applying Bayes' theorem.

Bayes' theorem [Bronstein et al., 2000] relates the joint probability $p(\mathbf{a}, \mathbf{b})$, conditional probabilities $p(\mathbf{a}|\mathbf{b})$ and $p(\mathbf{b}|\mathbf{a})$ as well as the prior probabilities $p(\mathbf{a})$ and $p(\mathbf{b})$ for two arbitrary random variables $\mathbf{a}$ and $\mathbf{b}$. It can be expressed by a multiplication of prior and conditional probabilities:

$$p(\mathbf{a}, \mathbf{b}) = p(\mathbf{a}|\mathbf{b}) \cdot p(\mathbf{b}) = p(\mathbf{b}|\mathbf{a}) \cdot p(\mathbf{a}). \tag{2.31}$$

The combination of (2.30) and (2.31) likewise describes the identification problem depending on the likelihood. The density function of the spoken phrase $p(\mathbf{x}_{1:T})$ is speaker independent and can be discarded. The resulting criterion is based on a multiplication of the likelihood, which describes the match between the speaker model and the observed data, and the prior probability for a particular speaker:

$$i_{\text{MAP}} = \arg\max_i \{p(\mathbf{x}_{1:T}|i) \cdot p(i)\}. \tag{2.32}$$

Depending on the application different approaches concerning the statistical models and training algorithms exist:

Speaker identification can be separated from *speaker verification* [Campbell, 1997]. Speaker identification has to recognize the voice of a current user without any prior knowledge about his identity whereas speaker verification accepts or rejects a claimed identity. Thus, speaker verification is typically applied to authentication problems [Campbell, 1997]. Speech plays an important role in biometric systems such as financial transactions or access control in security relevant areas [Campbell, 1997]. Speech can serve as biometric key to secure the access to computer networks and to ensure information security [Reynolds and Carlson, 1995]. The user enters an identity claim and is asked to speak a prompted text or a password [Campbell, 1997]. For example, speaker verification compares the observed voice characteristics with stored voice patterns of the claimed speaker identity and speech recognition checks the spoken password.

Speaker identification enables the personalization and optimization of speech controlled systems by accounting for user specific habits. Usability can be increased by speaker-adapted user-machine interfaces. Reynolds et al. [2000] provide an introduction into speaker verification which widely applies to speaker identification as well.

Speaker identification can be further subdivided into *closed-set* and *open-set* systems [Gish and Schmidt, 1994]. In closed-set scenarios the current user has to be assigned to one speaker out of a pool of enrolled speakers. The criterion for speaker selection can be given by the MAP criterion in (2.30). In open-set scenarios identification has not only to determine the speaker's identity but has also to decide whether the current user is an enrolled speaker or an unknown speaker.

A further discrimination is given by the general type of statistical model such as *non-parametric* and *parametric* models [Gish and Schmidt, 1994]. Non-parametric techniques usually admit any kind of distribution function to represent the measured distribution. Clustering algorithms such as k-means or Linde Buzo Gray (LBG) [Linde et al., 1980] are applied to capture speaker specific centers of gravity in the feature space. At run-time speaker identification algorithms, e.g. the *Vector Quantization* (VQ), rely on distance measures to find the speaker model with the highest match as found by Gish and Schmidt [1994].

Parametric approaches presume a special kind of statistical model a priori and only have to estimate the corresponding parameters. Gish and Schmidt [1994] provide a general description of different speaker identification techniques and refer to parametric and non-parametric identification.

Several strategies exist for the training of statistical models. The decision in (2.30) can be directly realized by *discriminative* speaker models [Bishop, 2007]. The goal is to learn how to separate enrolled speakers without the need to train individual statistical models for each speaker. However, training has to be re-computed when a new speaker is enrolled. The *Multilayer Perceptron* (MLP) as a special case of the *Neural Networks* (NN) framework[6] is a well-known discriminative model. For example, MLPs have been used for speaker identification as found by Genoud et al. [1999]; Morris et al. [2005]. Reynolds and Rose [1995] employ *Radial Basis Functions* (RBF) as a reference implementation for their proposed speaker identification framework based on generative models.

In contrast to discriminative models, *generative* speaker models individually represent the characteristics of the target speaker. They are independent from non-target speakers. The pool of enrolled speakers can be extended iteratively without re-training of all speaker models. The speaker whose model yields the best match with the observed feature vectors is considered as the target speaker. Reynolds et al. [2000]; Reynolds and Rose [1995] investigate the statistical modeling of speaker characteristics with Gaussian mixtures models which is one of the standard approaches for speaker identification based on generative speaker modeling.

The final design of a speaker identification implementation depends on the application. For verification the current user might be asked to read a prompted text so that the content of the utterance is known. Another example may be limited vocabulary, e.g. digits, in special applications. This constraint is known as *text-dependent* speaker identification or verification [Gish and Schmidt, 1994]. Kimball et al. [1997]; Reynolds and Carlson [1995] describe two approaches for text-dependent realizations to verify a speaker's identity. *Text-prompted* systems [Che et al., 1996; Furui, 2009], e.g. for access control, can ask the user to speak an utterance that is prompted on a display. This minimizes the risk that an impostor plays back an utterance via loudspeaker. A larger vocabulary has to be recognized but the transcription is still known. In contrast, *text-independent* identification has to handle arbitrary speech input [Gish and Schmidt, 1994; Kwon and Narayanan, 2005]. This identification task requires different statistical approaches.

The scope of this book requires a complete system characterized by a high flexibility to integrate new speakers and to handle individually trained speaker models. Users are not restricted to special applications or a given vocabulary so that speaker identification has to be operated in a text-independent mode. Furthermore, applications in an embedded system impose constraints concerning memory consumption and computational load. These requirements suggest a speaker identification approach based on generative speaker models. Thus, only the most dominant generative statistical model is introduced in Sect. 2.4.2. For a more detailed survey on standard techniques for speaker identification or verification, it is referred to Campbell [1997]; Gish and Schmidt [1994].

---

[6] Bishop [1996, 2007] provides an extensive investigation of the NN framework and MLPs, in particular.

2.4.2 Gaussian Mixture Models

Gaussian Mixture Models (GMMs) have emerged as the dominating generative statistical model in the state-of-the-art of speaker identification [Reynolds et al., 2000]. GMMs are attractive statistical models because they can represent various probability density functions if a sufficient number of parameters can be estimated [Reynolds et al., 2000].

GMMs comprise a set of N multivariate Gaussian density functions subsequently denoted by the index k. The resulting probability density function of a particular speaker model i is a convex combination of all density functions:

GMMs are constructed on standard multivariate Gaussian densities as given in (2.15) but introduce the component index k as a latent variable with the discrete probability $p(k|i)$. The weights

$$w_k^i = p(k|i) \tag{2.33}$$

characterize the prior contribution of the corresponding component to the density function of a GMM and fulfill the condition

$$\sum_{k=1}^{N} w_k^i = 1. \tag{2.34}$$

Each Gaussian density represents a conditional density function $p(\mathbf{x}_t|k,i)$. According to Bayes' theorem, the joint probability density function $p(\mathbf{x}_t,k|i)$ is given by the multiplication of both. The sum over all densities results in the multi-modal probability density of GMMs

$$p(\mathbf{x}_t|\Theta_i) = \sum_{k=1}^{N} p(k|\Theta_i) \cdot p(\mathbf{x}_t|k,\Theta_i) \tag{2.35}$$

$$= \sum_{k=1}^{N} w_k^i \cdot \mathcal{N}\left\{\mathbf{x}_t|\boldsymbol{\mu}_k^i, \boldsymbol{\Sigma}_k^i\right\}. \tag{2.36}$$

Each component density is completely determined by its mean vector $\boldsymbol{\mu}_k$ and covariance matrix $\boldsymbol{\Sigma}_k$. The parameter set

$$\Theta_i = \left\{w_1^i, \ldots, w_N^i, \boldsymbol{\mu}_1^i, \ldots, \boldsymbol{\mu}_N^i, \boldsymbol{\Sigma}_1^i, \ldots, \boldsymbol{\Sigma}_N^i\right\} \tag{2.37}$$

contains the weighting factors, mean vectors and covariance matrices for a particular speaker model i. $\mathbf{x}_t$ denotes the feature vector. For example, MFCC features or mean normalized MFCCs may be employed for speaker identification [Reynolds, 1995a; Reynolds et al., 2000]. Delta features may be used in addition [Reynolds et al., 2000].

Diagonal covariance matrices in contrast to full matrices are advantageous because of their computational efficiency [Reynolds et al., 2000]. Multivariate Gaussian distributions with diagonal covariance matrices comprise uncorrelated univariate distributions and even imply statistical independence [Hänsler, 2001]. The loss of information can be compensated by a larger set of multivariate Gaussian distributions so that correlated features can be accurately modeled by GMMs [Reynolds et al., 2000]. Because of the DCT a low correlation of the MFCC features can be assumed [Quatieri, 2002]. Finally, GMMs based on diagonal covariances have been found to outperform realizations with full covariance matrices [Quatieri, 2002; Reynolds et al., 2000].

Fig. 2.9 exemplifies the likelihood function for a GMM comprising 4 Gaussian distributions with diagonal covariance matrices.

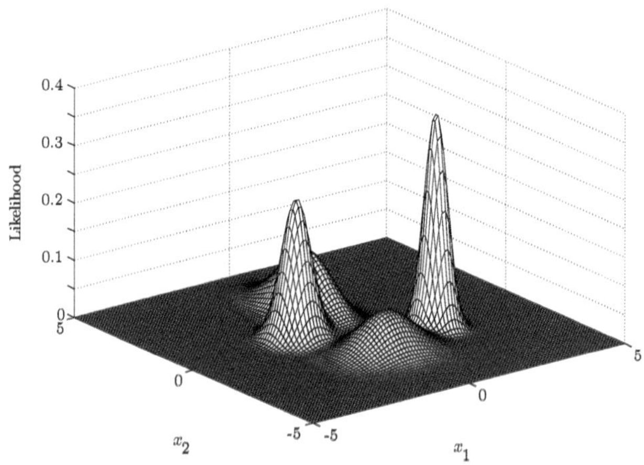

Fig. 2.9 Likelihood function of a GMM comprising 4 Gaussian densities. For convenience, two dimensional mean and feature vectors are chosen. x_1 and x_2 denote the elements of the feature vector.

The training of GMMs and speaker identification at run-time are described in the following.

Training

Subsequently, it is assumed that speaker specific data can be employed to train a GMM for each speaker. The goal of the training is to determine a parameter set Θ that optimally represents the voice characteristics of a particular speaker. For convenience, the speaker index is omitted.

If no prior knowledge about the parameter set is available, the optimization problem is to find a parameter set Θ_{ML} which maximizes the likelihood

$$\frac{\mathrm{d}}{\mathrm{d}\Theta} p(\mathbf{x}_{1:T}|\Theta)\bigg|_{\Theta=\Theta_{\mathrm{ML}}} = 0 \tag{2.38}$$

for a given training data set $\mathbf{x}_{1:T}$. Due to the latent variable k, the derivative of $p(\mathbf{x}_{1:T}|\Theta)$ cannot be solved analytically for its parameters as described in Sect. A.1.

The standard approximation is known as the *Expectation Maximization* (EM) algorithm [Dempster et al., 1977] which is characterized by an iterative procedure. Each iteration provides a new parameter set Θ starting from the parameter set of the preceding iteration $\bar{\Theta}$ or an initial set. The likelihood increases with each iteration until a maximum is reached. The likelihood acts as stopping condition. Each iteration consists of the following two steps:

- In the **E-step** all feature vectors are assigned to each Gaussian distribution by computing the posterior probability $p(k|\mathbf{x}_t, \bar{\Theta})$.
- In the **M-step** new ML estimates of the parameters are calculated based on the assignment of the E-step. An improved statistical modeling is therefore achieved by each iteration.

The EM algorithm is equivalently described by maximizing the auxiliary function

$$\begin{aligned} Q_{\mathrm{ML}}(\Theta, \bar{\Theta}) &= \mathrm{E}_{\mathrm{k}_{1:\mathrm{T}}}\{\log\left(p(\mathbf{x}_{1:T}, k_{1:T}|\Theta)\right)|\mathbf{x}_{1:T}, \bar{\Theta}\} \\ k_{1:T} &= \{k_1, \ldots, k_T\} \end{aligned} \tag{2.39}$$

with respect to Θ as found by Bishop [2007]; Gauvain and Lee [1994]. Under the iid assumption, maximizing

$$Q_{\mathrm{ML}}(\Theta, \bar{\Theta}) = \sum_{t=0}^{T-1} \sum_{k=0}^{N-1} p(k|\mathbf{x}_t, \bar{\Theta}) \cdot \log\left(p(\mathbf{x}_t, k|\Theta)\right) \tag{2.40}$$

leads to a locally optimal solution. However, it does not need to be the global optimum since the EM algorithm cannot distinguish between local and global maxima [Bishop, 2007]. Therefore different initialization strategies are described in the literature. Reynolds and Rose [1995] show that randomized initializations yield results comparable to more sophisticated strategies.

In the *E-step* each feature vector is assigned to all Gaussian densities by the posterior probability

$$p(k|\mathbf{x}_t, \bar{\Theta}) = \frac{\bar{w}_k \cdot \mathcal{N}\left\{\mathbf{x}_t|\bar{\boldsymbol{\mu}}_k, \bar{\boldsymbol{\Sigma}}_k\right\}}{\sum_{l=1}^{N} \bar{w}_l \cdot \mathcal{N}\left\{\mathbf{x}_t|\bar{\boldsymbol{\mu}}_l, \bar{\boldsymbol{\Sigma}}_l\right\}} \tag{2.41}$$

according to the importance for the training of a particular Gaussian density as found by Reynolds and Rose [1995]. The number of the softly assigned feature vectors is denoted by

$$n_k = \sum_{t=1}^{T} p(k|\mathbf{x}_t, \bar{\Theta}). \tag{2.42}$$

In the M-step the mean vectors, covariance matrices and weights are recalculated

$$\boldsymbol{\mu}_k^{\mathrm{ML}} = \frac{1}{n_k} \sum_{t=1}^{T} p(k|\mathbf{x}_t, \bar{\Theta}) \cdot \mathbf{x}_t \tag{2.43}$$

$$\boldsymbol{\Sigma}_k^{\mathrm{ML}} = \frac{1}{n_k} \sum_{t=1}^{T} p(k|\mathbf{x}_t, \bar{\Theta}) \cdot \mathbf{x}_t \cdot \mathbf{x}_t^T - \boldsymbol{\mu}_k^{\mathrm{ML}} \cdot (\boldsymbol{\mu}_k^{\mathrm{ML}})^T \tag{2.44}$$

$$w_k^{\mathrm{ML}} = \frac{n_k}{T} \tag{2.45}$$

based on the result of the E-step as found by Reynolds and Rose [1995]. If the covariance matrices are realized by diagonal matrices, only the main diagonal in (2.44) is retained. A lower bound should be assigned to avoid singularities in the likelihood function leading to infinite likelihood values [Bishop, 2007].

A more detailed description of GMMs and training algorithms can be found by Bishop [2007].

Evaluation

Speaker identification has to decide which user is speaking. Again, each utterance is characterized by a sequence of feature vectors $\mathbf{x}_{1:T}$ and statistical dependencies between successive time instances are neglected by the iid assumption. For each speaker i the log-likelihood

$$\log\left(p(\mathbf{x}_{1:T}|i)\right) = \sum_{t=1}^{T} \log\left(\sum_{k=1}^{N} w_k^i \cdot \mathcal{N}\left\{\mathbf{x}_t|\boldsymbol{\mu}_k^i, \boldsymbol{\Sigma}_k^i\right\}\right) \tag{2.46}$$

is calculated and the MAP criterion given by (2.32) is applied. The parameter set Θ is omitted. The speaker with the highest posterior probability $p(i|\mathbf{x}_{1:T})$ is identified. If the prior probability for the enrolled speakers is unknown or uniformly distributed, the ML criterion

$$i_{\mathrm{ML}} = \arg\max_i \left\{p(\mathbf{x}_{1:T}|i)\right\} \tag{2.47}$$

selects the speaker model with the highest match [Reynolds and Rose, 1995]. Under optimal conditions excellent speaker identification rates can be obtained as demonstrated by Reynolds [1995a].

Knowing the speaker identity enables speaker specific speech recognition as described in Sect. 3.3 or offers the possibility to further adapt the corresponding GMMs as explained in Sect. 2.6.

2.4.3 Detection of Unknown Speakers

The detection of unknown speakers is a critical issue for open-set speaker identification. A generative statistical model for a closed set of enrolled speakers was explained in Sect. 2.4.2. Each speaker participates in a training procedure in which speaker specific utterances are used to train speaker models. Detecting unknown or out-of-set speakers is difficult since unknown speakers cannot be modeled explicitly. Thus, another strategy is necessary to extend the GMM based speaker identification for out-of-set speakers.

A simple way is to introduce a threshold θ_{th} for the absolute log-likelihood values given by (2.46) as found by Fortuna et al. [2005]. The likelihood describes the match between a given statistical model and the observed data. If the speaker's identity does not correspond to a particular speaker model, a low likelihood value is expected. If all log-likelihoods of the enrolled speakers fall below a predetermined threshold

$$\log\left(p(\mathbf{x}_{1:T}|\Theta_i)\right) \leq \theta_{\text{th}}, \quad \forall i, \tag{2.48}$$

an unknown speaker has to be assumed.

However, in-car applications suffer from adverse environmental conditions such as engine noises. High fluctuations of the absolute likelihood are probable and may affect the threshold decision.

Advanced techniques may use normalization techniques comprising a *Universal Background Model* (UBM) as found by Angkititrakul and Hansen [2007]; Fortuna et al. [2005] or cohort models [Markov and Nakagawa, 1996]. The UBM is a speaker independent model which is trained on a large group of speakers. Instead of a threshold for log-likelihoods, the log-likelihood ratios of the speaker models and UBM can be examined for out-of-set detection [Fortuna et al., 2005]. If the following inequality

$$\log\left(p(\mathbf{x}_{1:T}|\Theta_i)\right) - \log\left(p(\mathbf{x}_{1:T}|\Theta_{\text{UBM}})\right) \leq \theta_{\text{th}}, \quad \forall i \tag{2.49}$$

is valid for all speaker models, an unknown speaker is likely. Alternatively, the ML criterion can be applied [Zhang et al., 2000].

The advanced approach has the advantage of lowering the influence of events that affect all statistical models in a similar way. For example, phrases spoken in an adverse environment may cause a mismatch between the speaker models and the audio signal due to background noises. Furthermore, text-dependent fluctuations in a spoken phrase, e.g. caused by unseen data or the training conditions, can be reduced [Fortuna et al., 2005; Markov and Nakagawa, 1996]. In those cases the log-likelihood ratio appears to be more robust than absolute likelihoods.

Speaker identification for known and unknown speakers can be implemented as a two-stage approach. In the first stage the most probable enrolled speaker is determined. Then the speaker identity is accepted or rejected by speaker verification based on the techniques described above [Angkititrakul and Hansen, 2007].

2.5 Speech Recognition

In the preceding sections several strategies were discussed to track different speakers. The algorithms presented try to recognize either a speaker turn or the speaker identity. Now the question has to be answered how to decode or recognize the understanding of a spoken utterance since the target system shall be operated by speech commands. In this context the influence of the speaker on speech recognition is temporarily neglected. Speaker variability will be explicitly addressed in the next section.

First, speech recognition is introduced from a general point of view and the problem of speech decoding is described mathematically. Then a statistical model is presented that has been established as a widely-used approach in *Automated Speech Recognition* (ASR) [Bishop, 2007]. Finally, the architecture of the speech recognizer used for the experiments of this book is explained. It is used in the following chapters. The basic components and the architecture of the speech recognizer are considered in more detail.

2.5.1 Motivation

For a speech controlled system it is essential to communicate with the user. On the one hand information has to be presented visually or acoustically to the user. On the other hand the user should be able to control the device via spoken commands. Hands-free telephony or the operation of a navigation system are in-car applications where speech recognition enables the user to speak a sequence of digits, select an entry from an address book or to enter city names, for example.

Speech recognition can be formulated as a MAP estimation problem. A word sequence $\boldsymbol{\mathcal{W}}_{1:N_{\mathrm{W}}} = \{\mathcal{W}_1, \mathcal{W}_2, \ldots, \mathcal{W}_{N_{\mathrm{W}}}\}$ has to be assigned to an utterance or equivalently a transcription of the spoken phrase has to be generated. N_{W} denotes the variable number of words within a single utterance. The most probable word sequence

$$\boldsymbol{\mathcal{W}}_{1:N_{\mathrm{W}}}^{\mathrm{MAP}} = \arg\max_{\boldsymbol{\mathcal{W}}_{1:N_{\mathrm{W}}}} \{p(\mathbf{x}_{1:T}|\boldsymbol{\mathcal{W}}_{1:N_{\mathrm{W}}}) \cdot p(\boldsymbol{\mathcal{W}}_{1:N_{\mathrm{W}}})\} \tag{2.50}$$

has to be determined out of all possible sequences [Schukat-Talamazzini, 1995; Setiawan et al., 2009]. The speech characteristics are represented by a sequence of feature vectors $\mathbf{x}_{1:T}$ which is extracted in the front end of the speech recognizer. For convenience, the LDA features $\mathbf{x}_t^{\mathrm{LDA}}$ introduced in Sect. 2.2 are used for speech recognition. The index LDA is omitted. The likelihood $p(\mathbf{x}_{1:T}|\boldsymbol{\mathcal{W}}_{1:N_{\mathrm{W}}})$ determines the match between the acoustic speech signal and the statistical model of a particular word sequence. The probability $p(\mathcal{W}_{1:N_{\mathrm{W}}})$ represents the speech recognizer's prior knowledge about given word sequences. This prior probability reflects how often these word sequences occur in a given language and whether this sequence is reasonable with respect to the grammar of a particular language model.

2.5.2 Hidden Markov Models

As discussed so far, speech characteristics and speaker variability were represented by particular time instances. Only the static features are tracked by GMMs. Statistical dependencies of successive time instances are usually not modeled because of the iid assumption. The dynamic of speech is therefore neglected.

Speech signals can be decomposed into phonemes as the smallest possible entity [Schukat-Talamazzini, 1995]. During speech production the vocal tract and the articulatory organs cannot change arbitrarily fast [O'Shaughnessy, 2000]. Phonemes can be defined by a characterizing sequence of states and transitions [O'Shaughnessy, 2000]. This becomes evident from Sect. 2.1 since vowels, nasal consonants, fricatives and plosives are produced by a specific configuration of the articulatory organs and excitation.

For example, 3 states can be used to represent the context to the prior phoneme, the steady-state of the actual phoneme and the context to the successor [O'Shaughnessy, 2000]. This motivates to integrate the dynamic behavior of speech into the statistical model.

GMMs can be extended by an underlying hidden statistical model to represent states and transitions. In a first step this underlying model is described and in a second step both models are combined. The result is the so-called *Hidden Markov Model* (HMM).

Markov Model

The Markov model is the statistical description of a system which can assume different states. It is characterized by a temporal limitation of the statistical dependencies [Hänsler, 2001]. Markov models are also called Markov models of first order if each transition only depends on the preceding and current state [Hänsler, 2001]. Statistical dependencies across multiple time instances are ignored.

A Markov model is defined by a set of states s, transition probabilities and an initialization. The joint probability $p(s_t, s_{t-1})$ is split by Bayes' theorem into a transition probability $p(s_t|s_{t-1})$ and the probability of the preceding state $p(s_{t-1})$. Finally, $p(s_t, s_{t-1})$ is reduced to the probability $p(s_t)$ by the sum over all previous states s_{t-1}:

$$p(s_t) = \sum_{s_{t-1}=1}^{M} p(s_t|s_{t-1}) \cdot p(s_{t-1}), \quad t > 1$$
$$a_{j_1 j_2} = p(s_t = j_2|s_{t-1} = j_1) \tag{2.51}$$

where M denotes the number of states. The transition probability is $a_{j_1 j_2}$ and the initial probability of a particular state is given by $p(s_1)$.

Fig. 2.10 displays a Markov chain as a feed-forward realization. The states are arranged in a chain and transitions can only occur either back to the original state or to one of the next states in a forward direction.

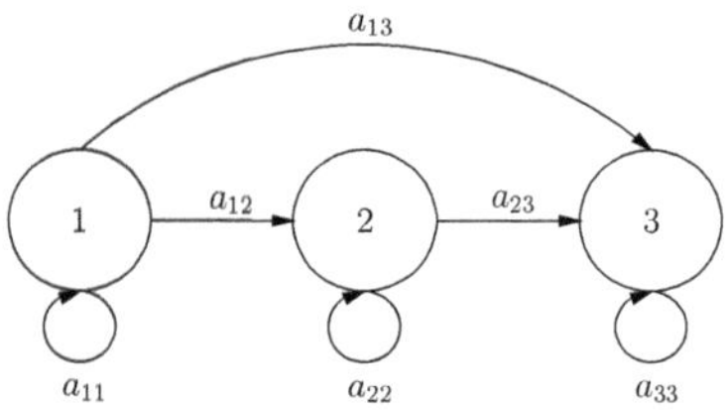

Fig. 2.10 Example for a Markov chain comprising 3 states organized as a feed-forward chain. The transition probabilities $a_{j_1 j_2}$ denote the transitions from state j_1 to state j_2.

Speech production of single phonemes can be modeled, e.g. by a feed-forward chain as given in Fig. 2.10. State 2 denotes the steady state of a phoneme whereas the states 1 and 3 represent the preceding and subsequent phoneme, respectively. However, variations of the pronunciation are not captured.

Fusion of Markov Model and GMM

An HMM fuses a Markov model and one or more GMMs. The Markov model represents the hidden unobserved random process and its states introduce the latent variable s_t. The emission probability represents the distribution of the observations belonging to the state s_t. It is modeled by a GMM which is subsequently called *codebook*. Bayes' theorem enables the combination of both random processes in analogy to Sect. 2.4.2. The final probability density function is composed by the superposition of all states of the Markov model and the Gaussian density functions of the GMM:

$$p(\mathbf{x}_t|\Theta) = \sum_{s_t=1}^{M} \sum_{k=1}^{N} p(\mathbf{x}_t|k, s_t, \Theta) \cdot p(k, s_t|\Theta) \tag{2.52}$$

$$= \sum_{s_t=1}^{M} \sum_{k=1}^{N} p(\mathbf{x}_t|k, s_t, \Theta) \cdot p(k|s_t, \Theta) \cdot p(s_t|\Theta). \tag{2.53}$$

N and k denote the number of Gaussian densities and the corresponding index. Θ contains all parameters of the HMM including the initial state probabilities, state transitions and the GMM parameters as defined in (2.37). For convenience, the following notation omits the parameter set.

The conditional discrete probability $p(k|s_t)$ corresponds to the weights of the GMM and the likelihood function $p(\mathbf{x}_t|k, s_t)$ is realized by a Gaussian density. The state probability $p(s_t)$ originates from the Markov model given by (2.51).

One realization is the *Semi-Continuous HMM* (SCHMM) [Huang and Jack, 1989; Rabiner and Juang, 1993; Schukat-Talamazzini, 1995]. It has only one GMM parameter set at its disposal. All states share the mean vectors and covariances of one GMM and only differ in their weighting factors:

$$p_{\mathrm{SCHMM}}(\mathbf{x}_t) = \sum_{s_t=1}^{M} \sum_{k=1}^{N} w_k^{s_t} \cdot \mathcal{N}\left\{\mathbf{x}_t|\boldsymbol{\mu}_k, \boldsymbol{\Sigma}_k\right\} \cdot p(s_t). \tag{2.54}$$

SCHMMs take benefit from the ability of GMMs to approximate arbitrary probability density functions. SCHMMs achieve high recognition accuracy. However, they are efficient in terms of memory and computational complexity [Huang and Jack, 1989; Schukat-Talamazzini, 1995]. The latter is essential especially for embedded systems. Furthermore, speaker adaptation is straightforward due to the possibility to define codebooks which can be easily modified [Rieck et al., 1992; Schukat-Talamazzini, 1995]. Speaker adaptation is usually realized by a shift of mean vectors whereas covariances are left unchanged as described in the next section. When using full covariance matrices, SCHMMs split the parameter set into mean vectors of low complexity and more complex covariances. This separation enables high speech recognition accuracies and makes adaptation efficient.

A further realization is given by the *Continuous Density HMM* (CDHMM) which supplies a complete GMM parameter set for each state:

$$p_{\mathrm{CDHMM}}(\mathbf{x}_t) = \sum_{s_t=1}^{M} \sum_{k=1}^{N} w_k^{s_t} \cdot \mathcal{N}\left\{\mathbf{x}_t|\boldsymbol{\mu}_k^{s_t}, \boldsymbol{\Sigma}_k^{s_t}\right\} \cdot p(s_t). \tag{2.55}$$

The superscript index s_t indicates the state dependence of the weight factor, mean vector and covariance matrix. CDHMMs enable an accurate statistical model for each state at the expense of a higher number of Gaussian distributions. The covariance matrices may be realized by diagonal matrices to limit the number of parameters.

Training and Evaluation

Speech can be modeled by HMMs. For instance, a phoneme can be represented by an HMM comprising 3 states. They reflect the steady state and the transitions from the preceding and to the subsequent phoneme within a spoken phrase. The corresponding GMMs represent the variability of the pronunciation.

The concatenation of models enables speech modeling on a word level. The corpus of all possible words is given by a lexicon. It determines the utterances to be recognized. The prior probabilities of word sequences are given by the language model [Schukat-Talamazzini, 1995].

HMMs can be trained by the Baum-Welch algorithm. In contrast to GMM training, state and transition probabilities have to be included. A detailed description can be found by Rabiner and Juang [1993]; Rabiner [1989].

The evaluation of HMMs has to determine the optimal sequence of models. Subsequently, the decoding problem is exemplified for only one HMM. The temporal sequence of states and transitions has to be determined so as to provide the highest match between model and the observed data.

The current state can be estimated by the forward algorithm [O'Shaughnessy, 2000; Rabiner and Juang, 1993; Schukat-Talamazzini, 1995] based on the feature vectors $\mathbf{x}_{1:t}$ observed so far and the initial state probability $p(s_1)$. The following notation is used: The density function $p(\mathbf{x}_{1:t}, s_t = j|\Theta)$ is denoted by $\alpha_t(j)$ for all states $j = 1, \ldots, M$. The conditional probability density function $p(\mathbf{x}_t|s_t = j)$ is given by $b_j(\mathbf{x}_t)$ and transitions of the Markov model are represented by $a_{lj} = p(s_t = j|s_{t-1} = l)$. The initialization is given by $\pi_j = p(s_1 = j)$. The forward algorithm can be iteratively calculated. The recursion is given by

$$\alpha_t(j) = b_j(\mathbf{x}_t) \cdot \sum_{l=1}^{M} \alpha_{t-1}(l) \cdot a_{lj}, \qquad t > 1 \tag{2.56}$$

and the initialization

$$\alpha_1(j) = \pi_j \cdot b_j(\mathbf{x}_1). \tag{2.57}$$

According to Bayes' theorem, the posterior of state s_t given the history of observations $\mathbf{x}_{1:t}$ can be obtained by a normalization of the forward algorithm:

$$p(s_t|\mathbf{x}_{1:t}, \Theta) = \frac{p(\mathbf{x}_{1:t}, s_t|\Theta)}{p(\mathbf{x}_{1:t}|\Theta)} \propto p(\mathbf{x}_{1:t}, s_t|\Theta). \tag{2.58}$$

In addition, the backward algorithm [O'Shaughnessy, 2000; Rabiner and Juang, 1993; Schukat-Talamazzini, 1995] can be applied when the complete utterance of length T is buffered. An algorithm similar to the forward algorithm is applied in reversed temporal order to calculate the likelihood $\beta_t(j) = p(\mathbf{x}_{t+1:T}|s_t = j, \Theta)$ for the successive observations $\mathbf{x}_{t+1:T}$ given the current state $s_t = j$. The recursion is given by

$$\beta_t(j) = \sum_{l=1}^{M} a_{jl} \cdot b_l(\mathbf{x}_{t+1}) \cdot \beta_{t+1}(l), \qquad 0 < t < T \tag{2.59}$$

and the initialization

$$\beta_T(j) = 1. \tag{2.60}$$

The posterior probability $p(s_t = j|\mathbf{x}_{1:T})$ can be calculated by combining the forward and backward algorithm

$$p(s_t = j|\mathbf{x}_{1:T}) = \frac{\alpha_t(j) \cdot \beta_t(j)}{\sum_{l=1}^{M} \alpha_t(l) \cdot \beta_t(l)}, \qquad 0 < t \leq T \tag{2.61}$$

as described by O'Shaughnessy [2000]; Rabiner and Juang [1993]; Schukat-Talamazzini [1995]. The most probable state

$$s_t^{\mathrm{MAP}} = \arg\max_{s_t}\{p(s_t|\mathbf{x}_{1:T})\} \tag{2.62}$$

can then be determined according to the MAP criterion.

To determine the best path in an HMM which means the most likely sequence of states, the Viterbi algorithm is usually employed [Forney, 1973; O'Shaughnessy, 2000; Rabiner and Juang, 1993; Schukat-Talamazzini, 1995]. In contrast to the forward algorithm, the density function $p(\mathbf{x}_{1:t}, s_{1:t}|\Theta)$ with $s_t = j$ has to be optimized:

$$\vartheta_t(j) = \max_{s_{t-1}} \{p(\mathbf{x}_{1:t}, s_{1:t}|\Theta)|s_t = j\}. \tag{2.63}$$

The Viterbi algorithm is recursively calculated by

$$\vartheta_t(j) = \max_{l} \{\vartheta_{t-1}(l) \cdot a_{lj} \cdot b_j(\mathbf{x}_t)\}, \quad t > 1 \tag{2.64}$$

$$\vartheta_1(j) = \pi_j \cdot b_j(\mathbf{x}_1). \tag{2.65}$$

Backtracking is required to determine the best sequence of states as described by Schukat-Talamazzini [1995]. The Viterbi algorithm is faster than the forward-backward algorithm since only the best sequence of states is considered instead of the sum over all paths [O'Shaughnessy, 2000].

For speech recognition as described by (2.50) this algorithm has to be extended so that transitions between HMMs can be tracked. Path search algorithms such as the Viterbi algorithm can be used to determine the recognition result given by the optimal word sequence $\boldsymbol{\mathcal{W}}_{1:N_{\mathrm{W}}}^{\mathrm{opt}} = \{\mathcal{W}_1^{\mathrm{opt}}, \ldots, \mathcal{W}_{N_{\mathrm{W}}}^{\mathrm{opt}}\}$. Further details about the evaluation of HMMs can be found by Rabiner and Juang [1993].

2.5.3 Implementation of an Automated Speech Recognizer

In this book, a speech recognizer based on SCHMMs is used. Fig. 2.11 shows the speech recognizer setup applied in the following chapters. It can be subdivided into three parts:

Fig. 2.11 Block diagram of the SCHMM-based speech recognizer used for the experiments of this book. A feature vector representation of the speech signal is extracted in the front-end of the speech recognizer. Each feature vector is compared with all Gaussian densities of the codebook. The result is used for speech decoding to generate a transcription of the spoken phrase. Figure is taken from [Herbig et al., 2010c].

- **Front-end.** In the front-end a noise reduction is performed, the MFCC features are extracted from the audio signal, a cepstral mean subtraction is calculated and an LDA is applied. For each time instance the vector representation $\mathbf{x}_t$ of the relevant speech characteristics is computed.
- **Codebook.** The feature vectors are evaluated by a speaker independent codebook subsequently called *standard codebook*. It consists of about 1000 multivariate Gaussian densities. The likelihood computation

$$p(\mathbf{x}_t|k) = \mathcal{N}\{\mathbf{x}|\boldsymbol{\mu}_k, \boldsymbol{\Sigma}_k\} \tag{2.66}$$

  of each Gaussian density with index k does not require a state alignment since only the weighting factors are state dependent. The soft quantization

$$\mathbf{q}_t = (q_1(\mathbf{x}_t), \ldots, q_k(\mathbf{x}_t), \ldots q_N(\mathbf{x}_t))^T \tag{2.67}$$

$$q_k(\mathbf{x}_t) = \frac{\mathcal{N}\{\mathbf{x}_t|\boldsymbol{\mu}_k, \boldsymbol{\Sigma}_k\}}{\sum_{l=1}^{N} \mathcal{N}\{\mathbf{x}_t|\boldsymbol{\mu}_l, \boldsymbol{\Sigma}_l\}} \tag{2.68}$$

  contains the matching result represented by the normalized likelihoods of all Gaussian densities [Fink, 2003]. The weights are evaluated in the speech decoder.
- **Speech Decoder.** The recognition task in (2.50) is solved by speech decoding based on $\mathbf{q}_{1:T}$. The acoustic model is realized by Markov chains. In addition, special models are used to capture garbage words and short speech pauses not removed by the speech segmentation. Garbages might contain speech which does not contribute to the speech recognition result. For example, a speaker hesitates within an utterance, mispronounces single words, coughs or clears his throat. In combination with the lexicon and language model, the transcription of the spoken utterance is obtained as depicted in Fig. 2.12. Utterances can be rejected when the confidence of the recognition result, e.g. given by the posterior probability $p(\mathcal{W}_{1:N_\mathrm{W}}|\mathbf{x}_{1:T})$, does not exceed a pre-determined threshold.

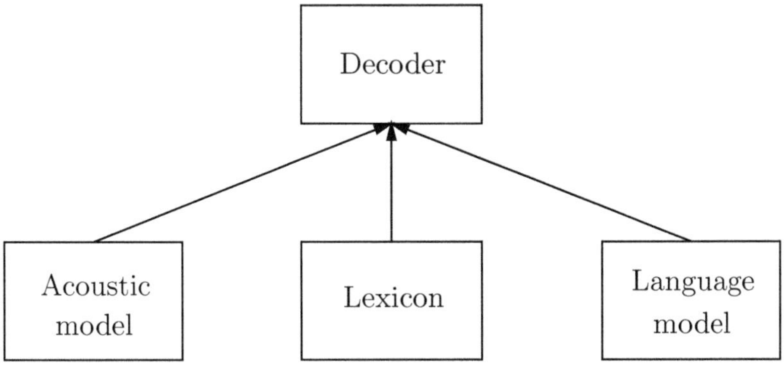

Fig. 2.12 Components of the speech decoding comprising the acoustic models, lexicon and language models.

2.6 Speaker Adaptation

In Sect. 2.2 speech production was discussed as a source-filter model. Speaker variability has its origins in a combination of gender and speaker specific excitation source, physical anatomy of the vocal tract and acquired speaking habits, e.g. speaking rate [Campbell, 1997]. Speech characteristics were also briefly described which are important for speech decoding.

In the preceding sections several techniques were introduced to deal with speaker and speech variability. In practical applications the problem of unseen conditions or time-variant speaker and speech characteristics occur. Since a re-training is usually not feasible at run-time, speaker adaptation provides an alternative method to adjust or initialize statistical models. Thus, speaker adaptation is an essential part of the speech controlled target system of this book.

In this section several adaptation strategies are discussed. They can be equally applied for speaker identification and speech recognition:

First, the application of speaker adaptation is motivated and possible sources of speech variations are given. The latter can affect speaker identification and speech recognition. Several applications for speaker adaptation in speaker identification and speech recognition are described. The differences between the adaptation of GMMs and HMMs are emphasized. A simplified notation is introduced to describe both adaptation scenarios.

Then two of the main representatives for long-term speaker adaptation are introduced. They enable the estimation of a high number of adaptation parameters. For each speaker an individually adapted statistical model can be given.

In contrast to this representative, an adaptation algorithm is described that excels in fast convergence. This algorithm is suitable for adaptation on limited training data since it only requires a few parameters. At the same time the minimal number of adaptation parameters restricts the algorithm's accuracy for long-term adaptation [Botterweck, 2001].

Finally, an adaptation strategy is introduced which integrates short-term and long-term adaptation capabilities. The optimal transition between short-term and long-term adaptation is achieved at the expense of a higher computational load [Jon et al., 2001; Stern and Lasry, 1987].

2.6.1 Motivation

In the preceding sections several statistical models and training algorithms were presented to handle speech and speaker related information. Speaker adaptation provides an alternative way to create adjusted statistical models.

The practical realization of the statistical models described in the preceding sections often suffers from mismatches between training and test conditions when the training has to be done on incomplete data that do not represent the actual test situation. Speaker adaptation offers the opportunity to moderate those mismatches and to obtain improved statistical representations.

In general, pattern recognition on speech encounters several sources which are responsible for a mismatch. Equation (2.4) reveals that the speaker's vocal tract and environmental influences may cause variations. Among those, there are the channel impulse response as well as background noises.

Even though the noise reduction of the front-end reduces background noises and smooths the power spectrum, the residual noise and channel distortions can affect the speaker identification and speech recognition accuracy.

Enhancement or normalization techniques, e.g. cepstral mean normalization, can be applied to reduce unwanted variations such as channel characteristics.

In the case of speaker specific variations speaker adaptation integrates them into the statistical modeling to initialize or enhance a speaker specific model. It starts from a preceding or initial parameter set $\bar{\Theta}$ and provides an optimized parameter set Θ for each speaker based on training data. The speaker index is omitted. Optimization can be achieved by the ML or MAP criterion which increase either the likelihood or posterior probability for the recorded utterance $\mathbf{x}_{1:T}$. The problem can be formulated by

$$\Theta_{\mathrm{ML}} = \arg\max_{\Theta} \{p(\mathbf{x}_{1:T}|\Theta)\} \tag{2.69}$$

$$\Theta_{\mathrm{MAP}} = \arg\max_{\Theta} \{p(\mathbf{x}_{1:T}|\Theta) \cdot p(\Theta)\} \tag{2.70}$$

as found by Gauvain and Lee [1994]. The same problem has to be solved as for the training algorithm in Sect. 2.4.2. Incomplete data have to be handled because of latent variables. This can be implemented by the EM algorithm as described in Sect. A.1.

In contrast to the realization of the EM algorithm in Sect. 2.4.2, just one iteration or only a few ones are performed. The first iterations usually contribute most to the final solution so that a reduction of the computational complexity is achieved by this restriction. Furthermore, the risk of over-fitting can be reduced. Equations (A.5) and (A.3) form the base of the following algorithms.

The capability of adaptation algorithms depends on the number of available parameters. With a higher number of parameters a more individual and accurate representation of the voice pattern of a particular speaker can be achieved. A high number of parameters, however, requires a large amount of adaptation data. Otherwise over-fitting can occur since the statistical model starts to represent insignificant properties and loses the ability to generalize [Duda et al., 2001].

Thus, either different approaches are needed for special applications depending on the amount of data or an integrated strategy has to be found.

2.6.2 Applications for Speaker Adaptation in Speaker Identification and Speech Recognition

GMMs were introduced in Sect. 2.4.2 as common statistical models to capture speaker characteristics. The former ones are trained by the EM algorithm to obtain an optimal statistical representation. However, in the literature there exist alternative methods to obtain speaker specific GMMs. According to Nishida and Kawahara [2005] adapting a Universal Background Model (UBM) to a particular speaker can be viewed as one of the most frequently applied techniques to build up speaker specific GMMs.

In the literature a further important application is updating the codebook of a speech recognizer [Zavaliagkos et al., 1995]. In Sect. 2.5 speech recognition was introduced from a speaker independent point of view. Many speech recognizers are extensively trained on a large group of speakers and thus do not represent the true speech characteristics of particular speakers [Zavaliagkos et al., 1995]. In realistic applications speaker adaptation can enhance the robustness of the ASR. The recognition rate can be significantly increased as demonstrated by Gauvain and Lee [1994]; Kuhn et al. [2000], for example.

The preceding sections showed that GMMs are a special case of HMMs. Therefore similar techniques can be applied to adapt GMMs and HMMs. Subsequently, only speech recognizers based on SCHMMs are investigated in this book.

Speaker adaptation for speaker identification purely aims at a better representation of the speaker's static voice characteristics with respect to the likelihood or posterior probability.

Codebook adaptation is intended to optimize the recognition of spoken phrases and therefore includes the temporal speech dynamics as a further aspect. Codebooks can be optimized by taking advantage from the speech recognizer's knowledge about the spoken utterance. Since speaker adaptation follows the speech decoding in the realization of this book, a two-stage approach becomes feasible. Hence, the optimization problem can be solved in analogy to the segmental MAP algorithm

$$\Theta_{\text{seg MAP}} = \arg\max_{\Theta} \left\{ \max_{s_{1:T}} \left\{ p(\mathbf{x}_{1:T}, s_{1:T}|\Theta) \cdot p(\Theta) \right\} \right\} \tag{2.71}$$

or equivalently by the following recursion

$$\hat{s}_{1:T} = \max_{s_{1:T}} \left\{ p(\mathbf{x}_{1:T}, s_{1:T}|\bar{\Theta}) \right\} \tag{2.72}$$

$$\hat{\Theta} = \arg\max_{\Theta} \left\{ p(\mathbf{x}_{1:T}, \hat{s}_{1:T}|\Theta) \cdot p(\Theta) \right\} \tag{2.73}$$

as found by Gauvain and Lee [1994]. The state sequence of an observed utterance is given by $s_{1:T} = \{s_1, s_2 \ldots, s_T\}$. $\bar{\Theta}$ and $\hat{\Theta}$ denote the parameter set of the preceding iteration and the adapted parameter set. Several iterations may be performed to calculate $\Theta_{\text{seg MAP}}$.

One problem of such a two-stage approach is the fact that transcription errors can negatively affect the speaker adaptation accuracy, especially when a large number of free parameters have to be estimated. Thus confidence measures may be employed on a frame or word level to increase the robustness against speech recognition errors [Gollan and Bacchiani, 2008].

The following approximation is used in this book. During speech decoding an optimal state sequence $\hat{s}_{1:T}$ is determined by the Viterbi algorithm. The speech recognition result is integrated in codebook optimization by the state dependent weights of the shared GMM since only SCHMMs are considered. The auxiliary functions of the EM algorithm in (A.3) and (A.5) have to be re-written

$$Q_{\text{seg MAP}}(\Theta, \bar{\Theta}) = Q_{\text{seg ML}}(\Theta, \bar{\Theta}) + \log\left(p(\Theta)\right) \tag{2.74}$$

$$Q_{\text{seg ML}}(\Theta, \bar{\Theta}) = \sum_{t=1}^{T} \sum_{k=1}^{N} p(k|\hat{s}_t, \mathbf{x}_t, \bar{\Theta}) \cdot \log\left(p(\mathbf{x}_t, k|\hat{s}_t, \Theta)\right) \tag{2.75}$$

$$\begin{aligned} &= \sum_{t=1}^{T} \sum_{k=1}^{N} p(k|\hat{s}_t, \mathbf{x}_t, \bar{\Theta}) \cdot \log\left(p(\mathbf{x}_t|k, \hat{s}_t, \Theta)\right) \\ &+ \sum_{t=1}^{T} \sum_{k=1}^{N} p(k|\hat{s}_t, \mathbf{x}_t, \bar{\Theta}) \cdot \log\left(p(k|\hat{s}_t, \Theta)\right) \end{aligned} \tag{2.76}$$

to integrate the knowledge of the speech transcription in the codebook adaptation.

When speaker adaptation is restricted to adapt only the mean vectors $\boldsymbol{\mu}_l$, $l = 1, \ldots, N$, the optimization of the auxiliary function can be simplified for SCHMMs:

$$\frac{\mathrm{d}}{\mathrm{d}\boldsymbol{\mu}_l} Q_{\mathrm{seg\,ML}}(\Theta, \bar{\Theta}) = \sum_{t=1}^{T} \sum_{k=1}^{N} p(k|\hat{s}_t, \mathbf{x}_t, \bar{\Theta}) \cdot \frac{\mathrm{d}}{\mathrm{d}\boldsymbol{\mu}_l} \log\left(p(\mathbf{x}_t|k, \Theta)\right). \quad (2.77)$$

The likelihood $p(\mathbf{x}_t|k, s_t, \Theta)$ of a particular Gaussian distribution does not depend on the current state s_t and the prior probability $p(k|s_t, \Theta) = w_k^{s_t}$ is independent from the mean vectors $\boldsymbol{\mu}_l$. The reader is referred to Gauvain and Lee [1994]; Matrouf et al. [2001] for adaptation scenarios modifying also the covariance matrices of an HMM.

The difference between the adaptation of GMMs and codebooks of SC-HMMs becomes now obvious. The posterior probability $p(k|\mathbf{x}_t, \bar{\Theta})$ determines the assignment of the observed feature vector $\mathbf{x}_t$ to the Gaussian distributions of a GMM. The posterior probability $p(k|\hat{s}_t, \mathbf{x}_t, \bar{\Theta})$ integrating the transcription of the spoken phrase is used for codebook adaptation. It provides more information about the speech signal compared to $p(k|\mathbf{x}_t, \bar{\Theta})$. Thus, the E-step does not only check whether a feature vector is assigned to a particular Gaussian distribution. In addition, the contribution to the recognition result gains importance for speaker adaptation.

Speaker adaptation normally modifies only the codebooks of a speech recognizer. Adapting the transitions of the Markov chain is usually discarded because the emission densities are considered to have a higher effect on the speech recognition rate compared to the transition probabilities [Dobler and Rühl, 1995]. The transition probabilities reflect the durational information [O'Shaughnessy, 2000]. For example, they contain the speech rate and average duration between the start and end state of a feed-forward chain as shown in Fig. 2.10. It is referred to Dobler and Rühl [1995]; Gauvain and Lee [1994] for adapting the Markov transitions.

Subsequently, the discussion is focused on GMM and codebook adaptation. Because of the two-staged approach a simplified notation can be used for both applications in speaker identification and speech recognition. The state variable is omitted in the case of a speech recognizer's codebook to obtain a compact representation. Thus, codebooks and GMMs can be equally considered when adaptation is described in this book.

2.6.3 Maximum A Posteriori

Maximum A Posteriori and Maximum Likelihood Linear Regression can be viewed as standard adaptation algorithms [Kuhn et al., 1999, 2001]. They both belong to the most frequently used techniques to establish speaker specific GMMs by adapting a UBM on speaker specific data [Nishida and Kawahara, 2005]. Codebook adaptation of a speech recognizer is a further important application as already discussed in Sect. 2.6.2.

MAP adaptation also known as Bayesian learning [Gauvain and Lee, 1994] is a standard approach. It is directly derived from the extended auxiliary function

$$Q_{\mathrm{MAP}}(\Theta, \bar{\Theta}) = Q_{\mathrm{ML}}(\Theta, \bar{\Theta}) + \log\left(p(\Theta)\right), \tag{2.78}$$

where $\bar{\Theta}$ denotes an initial or speaker specific parameter set. It integrates prior knowledge $p(\Theta)$ and the ML estimates Θ_{ML} based on the observed data [Stern and Lasry, 1987]. This algorithm is characterized by individual adaptation of each Gaussian density [Gauvain and Lee, 1994; Jon et al., 2001; Reynolds et al., 2000].

Gauvain and Lee [1994] provide a detailed derivation and the mathematical background of the Bayesian learning algorithm so that only the results are given here. More details can be found in Sect. A.2 where codebook adaptation is exemplified.

MAP adaptation starts from a speaker independent UBM as the initial parameter set Θ_{UBM} or alternatively from a trained speaker specific GMM. The parameters $\boldsymbol{\mu}_k^{\mathrm{UBM}}$, $\boldsymbol{\Sigma}_k^{\mathrm{UBM}}$ and w_k^{UBM} subsequently contain the prior knowledge about the mean vectors, covariance matrices and the weights, respectively.

Equation (2.78) has to be optimized with respect to the modified parameter set Θ. The feature vectors are viewed in the context of the iid assumption and statistical dependencies among the Gaussian densities are neglected. Thus, $p(\Theta)$ can be factorized [Zavaliagkos et al., 1995]. This allows individual adaptation equations for each Gaussian distribution. The following two steps are performed:

First, a new parameter set of ML estimates Θ_{ML} is calculated by applying one E-step and one M-step of the EM algorithm. The ML estimates are denoted by the superscript ML. N and n_k denote the total number of Gaussian densities and the number of feature vectors softly assigned to a particular Gaussian density. For convenience, the equations of the ML estimation in Sect. 2.4 are modified

$$p(k|\mathbf{x}_t, \Theta_{\mathrm{UBM}}) = \frac{w_k^{\mathrm{UBM}} \cdot \mathcal{N}\left\{\mathbf{x}_t | \boldsymbol{\mu}_k^{\mathrm{UBM}}, \boldsymbol{\Sigma}_k^{\mathrm{UBM}}\right\}}{\sum_{l=1}^{N} w_l^{\mathrm{UBM}} \cdot \mathcal{N}\left\{\mathbf{x}_t | \boldsymbol{\mu}_l^{\mathrm{UBM}}, \boldsymbol{\Sigma}_l^{\mathrm{UBM}}\right\}} \tag{2.79}$$

$$n_k = \sum_{t=1}^{T} p(k|\mathbf{x}_t, \Theta_{\mathrm{UBM}}) \tag{2.80}$$

$$\boldsymbol{\mu}_k^{\mathrm{ML}} = \frac{1}{n_k} \sum_{t=1}^{T} p(k|\mathbf{x}_t, \Theta_{\mathrm{UBM}}) \cdot \mathbf{x}_t \tag{2.81}$$

$$\boldsymbol{\Sigma}_k^{\mathrm{ML}} = \frac{1}{n_k} \sum_{t=1}^{T} p(k|\mathbf{x}_t, \Theta_{\mathrm{UBM}}) \cdot (\mathbf{x}_t - \boldsymbol{\mu}_k^{\mathrm{ML}}) \cdot (\mathbf{x}_t - \boldsymbol{\mu}_k^{\mathrm{ML}})^T \tag{2.82}$$

$$w_k^{\mathrm{ML}} = \frac{n_k}{T} \tag{2.83}$$

so that the UBM determines the feature vector assignment and acts as the initial model. When an iterative procedure is applied, Θ_{UBM} has to be replaced by $\bar{\Theta}$ which represents the parameter set of the preceding iteration.

After the ML estimation two sets of parameters exist. The ML estimates Θ_{ML} represent the observed data whereas the initial parameters Θ_{UBM} comprise the prior distribution. For adaptation a smooth transition between Θ_{UBM} and Θ_{ML} has to be found depending on these observations.

This problem is solved in the second step by a convex combination of the initial parameters and the ML estimates resulting in the optimized parameter set Θ_{MAP}. The number of softly assigned feature vectors n_k controls for each Gaussian density the weighting of this convex combination. The influence of the new estimates Θ_{ML} increases with higher n_k. Small values of n_k cause the MAP estimate to resemble the prior parameters. The associated equations

$$\alpha_k = \frac{n_k}{n_k + \eta} \tag{2.84}$$

$$\boldsymbol{\mu}_k^{\mathrm{MAP}} = (1 - \alpha_k) \cdot \boldsymbol{\mu}_k^{\mathrm{UBM}} + \alpha_k \cdot \boldsymbol{\mu}_k^{\mathrm{ML}} \tag{2.85}$$

$$\boldsymbol{\Sigma}_k^{\mathrm{MAP}} = (1 - \alpha_k) \cdot \boldsymbol{\Sigma}_k^{\mathrm{UBM}} + \alpha_k \cdot \boldsymbol{\Sigma}_k^{\mathrm{ML}} \tag{2.86}$$

$$w_k^{\mathrm{MAP}} \propto (1 - \alpha_k) \cdot w_k^{\mathrm{UBM}} + \alpha_k \cdot w_k^{\mathrm{ML}}, \qquad \sum_{l=1}^{N} w_l^{\mathrm{MAP}} = 1 \tag{2.87}$$

can be derived from (2.78) and can be found by Reynolds et al. [2000].

The constant η contains prior knowledge about the densities of the GMM [Gauvain and Lee, 1994]. In practical applications η can be set irrespectively of this. Reynolds et al. [2000] show that the choice of this constant η is not sensitive for the performance of speaker verification and can be selected in a relatively wide range. The weight computation w_k^{MAP} introduces a constant factor to assure that the resulting weights are valid probabilities and sum up to unity.

Adapting the mean vectors seems to have the major effect on the speaker verification results [Reynolds et al., 2000]. Thus, only mean vectors and if applicable weights can be adapted as a compromise between enhanced identification rate and computational load.

Fig. 2.13(a) and Fig. 2.13(b) illustrate the procedure of the MAP adaptation in analogy to Reynolds et al. [2000]: Fig. 2.13(a) displays the original GMM as ellipses. The centers represent mean vectors and the main axes of the ellipses correspond to variances. The crosses indicate the observed feature vectors. Fig. 2.13(b) shows the adapted Gaussian densities as filled ellipses and the original density functions as unfilled ellipses. The Gaussian densities at the top are individually adapted with respect to the assigned feature vectors. The resulting ellipses cover the feature vectors and parts of the original ellipses indicating the convex combination in (2.85). One density function remains unchanged since no feature vectors are assigned.

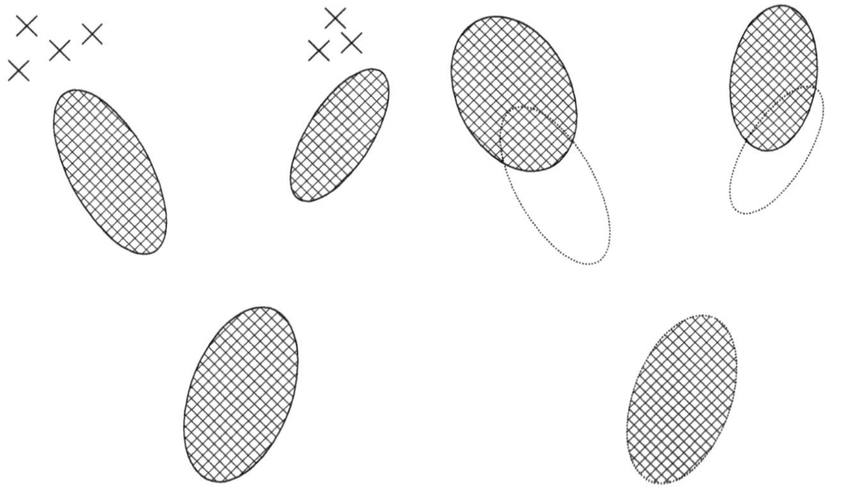

(a) Filled ellipses display the original Gaussian densities of a GMM. The observed feature vectors are depicted as crosses.

(b) Adapted Gaussian densities are depicted as filled ellipses whereas the initial Gaussian densities of the GMM are denoted by unfilled ellipses (dotted line).

Fig. 2.13 Example for the MAP adaptation of a GMM comprising 3 Gaussian densities.

Finally, the advantages and drawbacks of the MAP adaptation should be discussed: Individual adaptation is achieved on extensive adaptation data which enables high adaptation accuracy. If the constant η is selected appropriately, MAP adaptation is robust against over-fitting caused by limited speaker specific data. However, inefficient adaptation of untrained Gaussian densities has to be expected resulting in slow convergence [Zavaliagkos et al., 1995]. For $n_k \to 0$ no adaptation occurs or only a few parameters are marginally adapted. Thus, MAP adaptation is important for long-term adaptation and suffers from inefficient adaptation on limited data [Kuhn et al., 2000].

2.6.4 Maximum Likelihood Linear Regression

Approaches for speaker adaptation such as MAP adaptation, where each Gaussian distribution is adapted individually, require a sufficient amount of training data to be efficient. In general the number of adaptation parameters has to be balanced with the amount of speaker specific data.

One way is to estimate a set of model transformations for classes of Gaussian distributions to capture speaker characteristics. Several publications such as Gales and Woodland [1996]; Leggetter and Woodland [1995a] describe *Maximum Likelihood Linear Regression* (MLLR) which is widely used in speaker adaptation. The key idea of MLLR is to represent model adaptation by a linear transformation. For example, mean vector adaptation may be realized for each regression class r by a multiplication of the original mean vector $\boldsymbol{\mu}_{k_r}$ with matrix A_r and an offset vector $\mathbf{b}_r$. The optimized mean vector $\boldsymbol{\mu}_{k_r}^{\mathrm{MLLR}}$ is given by

$$\boldsymbol{\mu}_{k_r}^{\mathrm{MLLR}} = \mathsf{A}_r^{\mathrm{opt}} \cdot \boldsymbol{\mu}_{k_r} + \mathbf{b}_r^{\mathrm{opt}} \tag{2.88}$$

which can be rewritten as

$$\boldsymbol{\mu}_{k_r}^{\mathrm{MLLR}} = \mathsf{W}_r^{\mathrm{opt}} \cdot \boldsymbol{\zeta}_{k_r} \tag{2.89}$$

$$\mathsf{W}_r^{\mathrm{opt}} = \left[\mathbf{b}_r^{\mathrm{opt}} \ \mathsf{A}_r^{\mathrm{opt}}\right] \tag{2.90}$$

$$\boldsymbol{\zeta}_{k_r}^T = \left[\mathbf{1}^T \ \boldsymbol{\mu}_{k_r}^T\right]. \tag{2.91}$$

The covariance matrices may be adapted in a similar way as described by Gales and Woodland [1996].

The goal of MLLR is to determine an optimal transformation matrix $\mathsf{W}_{k_r}^{\mathrm{opt}}$ in such a way that the new parameter set maximizes the likelihood of the training data. This problem may be solved by maximizing the auxiliary function $Q_{\mathrm{ML}}(\Theta, \bar{\Theta})$ of the EM algorithm. In this context $\bar{\Theta}$ denotes the initial parameter set, e.g. the standard codebook Θ_0, or the parameter set of the previous iteration when an iterative procedure is applied. The optimal transformation matrix $\mathsf{W}_{k_r}^{\mathrm{opt}}$ is given by the solution of the following set of equations:

$$\sum_{t=1}^{T}\sum_{r=1}^{R} p(k_r|\mathbf{x}_t, \bar{\Theta}) \cdot \Sigma_{k_r}^{-1} \cdot \mathbf{x}_t \cdot \boldsymbol{\zeta}_{k_r}^T = \sum_{t=1}^{T}\sum_{r=1}^{R} p(k_r|\mathbf{x}_t, \bar{\Theta}) \cdot \Sigma_{k_r}^{-1} \cdot \mathsf{W}_m^{\mathrm{opt}} \cdot \boldsymbol{\zeta}_{k_r} \cdot \boldsymbol{\zeta}_{k_r}^T \tag{2.92}$$

as found by Gales and Woodland [1996]. $p(k_r|\mathbf{x}_t, \bar{\Theta})$ denotes the posterior probability where each feature vector $\mathbf{x}_t$ is assigned to a particular Gaussian density k_r within a given regression class r.

In this book, MAP adaptation is viewed as an appropriate candidate for long-term adaptation. A highly specific speaker modeling can be achieved on extensive data. Furthermore, the combination of short-term and long-term adaptation in Sect. 4.2 can be intuitively explained. Thus, MLLR is not considered.

2.6.5 Eigenvoices

The MAP adaptation has its strength in long-term speaker adaptation whereas the *Eigenvoice* (EV) approach is advantageous in the case of limited training data as shown by Kuhn et al. [2000]. This technique benefits from prior knowledge about the statistical dependencies between all Gaussian densities. Thus, this algorithm significantly reduces the number of adaptation parameters. Even very few training data, e.g. short command and control utterances, allows estimating some scalar parameters. When prior knowledge about speaker variability can be employed, the codebook of a speech recognizer can be efficiently adapted by some 10 adaptation parameters [Kuhn et al., 2000].

There are about 25,000 parameters to be adapted when the MAP adaptation is applied to the codebook of the speech recognizer introduced in Sect. 2.5.3. The EV technique in the realization of this book only needs 10 parameters to efficiently adapt codebooks. GMMs for speaker identification can be realized by significantly fewer parameters as will be shown in Sect. 5.3.

In an off-line training step codebook adaptation is performed for a large pool of speakers. The main directions of variations in the feature space are extracted. They are presented by eigenvectors which are subsequently called eigenvoices. At run-time codebook adaptation can be efficiently implemented by a linear combination of all eigenvoices. Only the weighting factors have to be determined to modify the model parameters in an optimal manner. For convenience, this presentation only considers the mean vector adaptation of codebooks but may be extended to weights or covariance matrices. A more detailed investigation of the EV approach can be found by Kuhn et al. [2000]; Thyes et al. [2000].

Training

In an off-line training step N_{Sp} speaker specific codebooks are trained. The origin of all speaker specific models is given here by the standard codebook with parameter set Θ_0. Each speaker with index i provides a large amount of training utterances so that an efficient long-term speaker adaptation is possible. For convenience, the MAP algorithm introduced in Sect. 2.6.3 is considered in the following. The result is a speaker specific set of mean vectors

$$\boldsymbol{\mu}_k^0 \overset{\mathrm{MAP}}{\longrightarrow} \left\{\boldsymbol{\mu}_k^1, \ldots, \boldsymbol{\mu}_k^i, \ldots, \boldsymbol{\mu}_k^{N_{\mathrm{Sp}}}\right\}, \qquad k = 1, \ldots, N \tag{2.93}$$

which is grouped by the component index k. In total, N Gaussian densities are employed.

Subsequently, all mean vectors are arranged in supervectors

$$\check{\boldsymbol{\mu}}^i = \begin{pmatrix} \boldsymbol{\mu}_1^i \\ \vdots \\ \boldsymbol{\mu}_k^i \\ \vdots \\ \boldsymbol{\mu}_N^i \end{pmatrix} \tag{2.94}$$

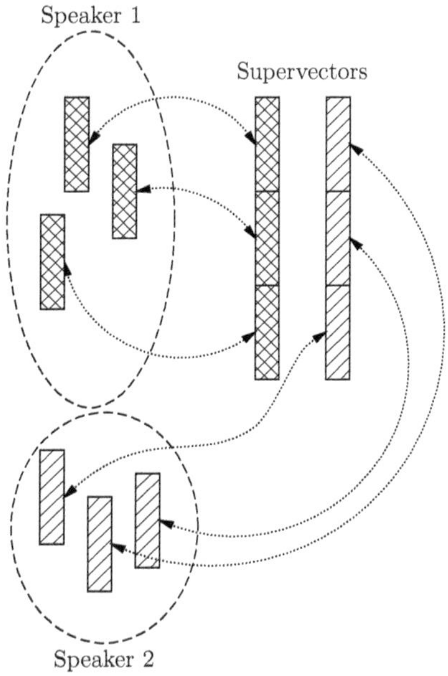

Fig. 2.14 Example for the supervector notation. All mean vectors of a codebook are stacked into a long vector.

indicated by the superscript in $\check{\boldsymbol{\mu}}$ and the missing index k. To obtain a compact notation, equation (2.93) is represented by supervectors

$$\check{\boldsymbol{\mu}}^0 \stackrel{\text{MAP}}{\longrightarrow} \left\{\check{\boldsymbol{\mu}}^1, \check{\boldsymbol{\mu}}^2, \ldots, \check{\boldsymbol{\mu}}^{N_{\mathrm{Sp}}}\right\}. \tag{2.95}$$

An example of the supervector notation is displayed in Fig 2.14. In the following, a pool of diverse speakers is considered. For convenience, the mean over all speaker specific supervectors is assumed to be identical to the supervector of the speaker independent mean vectors given by the standard codebook

$$\mathrm{E}_{\mathrm{i}}\{\check{\boldsymbol{\mu}}^i\} = \check{\boldsymbol{\mu}}^0. \tag{2.96}$$

An additional offset vector is omitted here.

Now the training has to obtain the most important adaptation directions which can be observed in this speaker pool. The covariance matrix

$$\check{\boldsymbol{\Sigma}}^{\mathrm{EV}} = \mathrm{E}_{\mathrm{i}}\{(\check{\boldsymbol{\mu}}^i - \check{\boldsymbol{\mu}}^0) \cdot (\check{\boldsymbol{\mu}}^i - \check{\boldsymbol{\mu}}^0)^T\} \tag{2.97}$$

is calculated and the eigenvectors of $\check{\boldsymbol{\Sigma}}^{\mathrm{EV}}$ are extracted using PCA[7] as described by Jolliffe [2002]. The eigenvalues are denoted by λ^{EV}. The corresponding eigenvectors $\check{\mathbf{e}}^{\mathrm{EV}}$ fulfill the condition

$$\check{\boldsymbol{\Sigma}}^{\mathrm{EV}} \cdot \check{\mathbf{e}}_l^{\mathrm{EV}} = \lambda^{\mathrm{EV}} \cdot \check{\mathbf{e}}_l^{\mathrm{EV}}, \qquad 1 \leq l \leq N_{\mathrm{Sp}}. \tag{2.98}$$

and can be chosen as orthonormal

$$(\check{\mathbf{e}}_{l_1}^{\mathrm{EV}})^T \cdot \check{\mathbf{e}}_{l_2}^{\mathrm{EV}} = \delta_{\mathrm{K}}(l_1, l_2), \quad 1 \leq l_1, l_2 \leq N_{\mathrm{Sp}} \tag{2.99}$$

since the covariance matrix is symmetric $\check{\boldsymbol{\Sigma}}^{\mathrm{EV}} = (\check{\boldsymbol{\Sigma}}^{\mathrm{EV}})^T$ by its definition [Bronstein et al., 2000; Meyberg and Vachenauer, 2003]. The number of eigenvoices is limited by the number of speakers N_{Sp}.

The eigenvalues characterize the variance of the observed data projected onto these eigenvectors if the eigenvectors are normalized [Jolliffe, 2002]. High values indicate the important directions in the sense of an efficient speaker adaptation. The eigenvectors $\check{\mathbf{e}}^{\mathrm{EV}}$ are sorted along decreasing eigenvalues and the following considerations focus only on the first L eigenvectors.

The first eigenvector usually represents gender information and enables gender detection as the results of Kuhn et al. [2000] suggest. This book makes use of only $L = 10$ eigenvoices similar to Kuhn et al. [2000] because the corresponding eigenvalues diminish rapidly.

Up to this point the principle directions of speaker adaptation are represented as supervectors. At run-time the EV speaker adaptation and especially the combination of short-term and long-term adaptation in Sect. 4.2 become more feasible when each Gaussian density is individually represented. Each supervector $\check{\mathbf{e}}_l^{\mathrm{EV}}$ separates into its elements

$$\check{\mathbf{e}}_l^{\mathrm{EV}} = \begin{pmatrix} \mathbf{e}_{1,l}^{\mathrm{EV}} \\ \vdots \\ \mathbf{e}_{k,l}^{\mathrm{EV}} \\ \vdots \\ \mathbf{e}_{N,l}^{\mathrm{EV}} \end{pmatrix}, \qquad l = 1, \ldots, L \tag{2.100}$$

where the index k indicates the assignment to a particular Gaussian density.

Adaptation

At run-time the optimal combination of the eigenvoices has to be determined to obtain a set of adapted parameters. The adapted mean vector $\check{\boldsymbol{\mu}}^{\mathrm{EV}}$ results from a linear combination of the original speaker independent mean vector $\check{\boldsymbol{\mu}}^0$ and the eigenvoices $\check{\mathbf{e}}^{\mathrm{EV}}$. For convenience, the representation of the eigenvoices for each Gaussian density is used

[7] Realizations of the PCA which are based on the correlation matrix also exist [Jolliffe, 2002; Kuhn et al., 2000].

$$\boldsymbol{\mu}_k^{\mathrm{EV}} = \boldsymbol{\mu}_k^0 + \sum_{l=1}^{L} \alpha_l \cdot \mathbf{e}_{k,l}^{\mathrm{EV}} \tag{2.101}$$

as noted in (2.100). In order to obtain a comprehensive notation the matrix

$$\mathsf{M}_k = (\mathbf{e}_{k,1}^{\mathrm{EV}}, \ldots, \mathbf{e}_{k,L}^{\mathrm{EV}}), \qquad k = 1, \ldots N \tag{2.102}$$

containing all eigenvoices is introduced. The weight vector $\boldsymbol{\alpha} = (\alpha_1, \ldots, \alpha_L)^T$ is employed to represent (2.101) by a matrix product and an offset vector:

$$\boldsymbol{\mu}_k^{\mathrm{EV}} = \boldsymbol{\mu}_k^0 + \mathsf{M}_k \cdot \boldsymbol{\alpha}. \tag{2.103}$$

For adaptation the optimal scalar weighting factors α_l have to be determined. Standard approaches solve this problem by optimizing the auxiliary function of the EM algorithm

$$Q_{\mathrm{ML}}(\Theta, \Theta_0) = \sum_{t=1}^{T} \sum_{k=1}^{N} p(k|\mathbf{x}_t, \Theta_0) \cdot [\log(p(\mathbf{x}_t|k, \Theta)) + \log(p(k|\Theta))] \tag{2.104}$$

as demonstrated by Kuhn et al. [1998], for example. In this context, Θ contains the weighting factors w_k^0, the mean vectors $\boldsymbol{\mu}_k^{\mathrm{EV}}$ depending on $\boldsymbol{\alpha}$ and the covariance matrices $\boldsymbol{\Sigma}_k^0$. w_k^0 and $\boldsymbol{\Sigma}_k^0$ are given by the standard codebook. When speaker adaptation is iteratively performed, Θ_0 has to be replaced by the parameter set $\bar{\Theta}$ of the preceding iteration. Prior knowledge can be included by optimizing

$$Q_{\mathrm{MAP}}(\Theta, \Theta_0) = Q_{\mathrm{ML}}(\Theta, \Theta_0) + \log(p(\Theta)). \tag{2.105}$$

$p(\mathbf{x}_t|k, \Theta)$ and $p(k|\Theta)$ were already introduced in Sect. 2.4.2. They denote the likelihood function of a single Gaussian density and the weight of a particular density function as given by the equations (2.35) and (2.36).

The derivations known from literature differ in the choice of the prior knowledge $p(\Theta)$ concerning the parameter set as found by Huang et al. [2004]; Kuhn et al. [1998], for example. In this book a uniform distribution is chosen so that the MAP criterion reduces to the ML estimation problem.

The optimal combination of the eigenvoices is obtained by inserting the linear combination (2.103) into the auxiliary function (2.104) and optimizing $Q_{\mathrm{ML}}(\Theta, \Theta_0)$ with respect to the weighting factors $\boldsymbol{\alpha}$. The computation is simplified since the weights are independent of the mean vectors:

$$\frac{\mathrm{d}}{\mathrm{d}\boldsymbol{\alpha}} \log(p(k|\Theta)) = 0 \tag{2.106}$$

$$\frac{\mathrm{d}}{\mathrm{d}\boldsymbol{\alpha}} \log(p(\mathbf{x}_t|k, \Theta)) = \frac{\mathrm{d}}{\mathrm{d}\boldsymbol{\alpha}} \log\left(\mathcal{N}\left\{\mathbf{x}_t|\boldsymbol{\mu}_k^{\mathrm{EV}}(\boldsymbol{\alpha}), \boldsymbol{\Sigma}_k^0\right\}\right). \tag{2.107}$$

Now the problem of the parameter computation can be given in a compact notation

$$\frac{\mathrm{d}}{\mathrm{d}\boldsymbol{\alpha}} Q_{\mathrm{ML}}(\Theta, \Theta_0) = -\frac{1}{2} \cdot \sum_{t=0}^{T-1} \sum_{k=1}^{N} p(k|\mathbf{x}_t, \Theta_0) \cdot \frac{\mathrm{d}}{\mathrm{d}\boldsymbol{\alpha}} d_k^{\mathrm{Mahal}}(\boldsymbol{\alpha}). \tag{2.108}$$

The squared Mahalanobis distance

$$d_k^{\mathrm{Mahal}}(\boldsymbol{\alpha}) = \left(\mathbf{x}_t - \boldsymbol{\mu}_k^{\mathrm{EV}}(\boldsymbol{\alpha})\right)^T \cdot (\boldsymbol{\Sigma}_k^0)^{-1} \cdot \left(\mathbf{x}_t - \boldsymbol{\mu}_k^{\mathrm{EV}}(\boldsymbol{\alpha})\right) \tag{2.109}$$

is given by the exponent of the Gaussian density excluding the scaling factor $-\frac{1}{2}$ as found by Campbell [1997]. d_k^{Mahal} is a quadratic distance measure and therefore the derivative leads to linear expressions in terms of $\boldsymbol{\alpha}$. Since $\boldsymbol{\Sigma}_k^0$ is symmetric, the derivation rule

$$\begin{aligned}
&\frac{\mathrm{d}}{\mathrm{d}\boldsymbol{\alpha}} d_k^{\mathrm{Mahal}}(\boldsymbol{\alpha}) \\
=&\frac{\mathrm{d}}{\mathrm{d}\boldsymbol{\alpha}} \left(\mathbf{x}_t - \boldsymbol{\mu}_k^0 - \mathsf{M}_k \cdot \boldsymbol{\alpha}\right)^T \cdot (\boldsymbol{\Sigma}_k^0)^{-1} \cdot \left(\mathbf{x}_t - \boldsymbol{\mu}_k^0 - \mathsf{M}_k \cdot \boldsymbol{\alpha}\right) && (2.110)\\
=&\frac{\mathrm{d}}{\mathrm{d}\boldsymbol{\alpha}} \boldsymbol{\alpha}^T \cdot \mathsf{M}_k^T \cdot (\boldsymbol{\Sigma}_k^0)^{-1} \cdot \mathsf{M}_k \cdot \boldsymbol{\alpha} - \frac{\mathrm{d}}{\mathrm{d}\boldsymbol{\alpha}} \boldsymbol{\alpha}^T \cdot \mathsf{M}_k^T \cdot (\boldsymbol{\Sigma}_k^0)^{-1} \cdot \left(\mathbf{x}_t - \boldsymbol{\mu}_k^0\right) \\
&- \frac{\mathrm{d}}{\mathrm{d}\boldsymbol{\alpha}} \left(\mathbf{x}_t - \boldsymbol{\mu}_k^0\right)^T \cdot (\boldsymbol{\Sigma}_k^0)^{-1} \cdot \mathsf{M}_k \cdot \boldsymbol{\alpha} && (2.111)\\
=&2 \cdot \mathsf{M}_k^T \cdot (\boldsymbol{\Sigma}_k^0)^{-1} \cdot \mathsf{M}_k \cdot \boldsymbol{\alpha} - 2 \cdot \mathsf{M}_k^T \cdot (\boldsymbol{\Sigma}_k^0)^{-1} \cdot \left(\mathbf{x}_t - \boldsymbol{\mu}_k^0\right) && (2.112)\\
=&-2 \cdot (\mathsf{M}_k)^T \cdot (\boldsymbol{\Sigma}_k^0)^{-1} \cdot \left(\mathbf{x}_t - \boldsymbol{\mu}_k^0 - \mathsf{M}_k \cdot \boldsymbol{\alpha}\right) && (2.113)
\end{aligned}$$

is applicable as found by Felippa [2004]; Petersen and Pedersen [2008].

The optimal parameter set $\boldsymbol{\alpha}^{\mathrm{ML}}$ has to fulfill the condition

$$\left.\frac{\mathrm{d}}{\mathrm{d}\boldsymbol{\alpha}} Q_{\mathrm{ML}}\right|_{\boldsymbol{\alpha}=\boldsymbol{\alpha}^{\mathrm{ML}}} = \mathbf{0} \tag{2.114}$$

which leads to the following set of equations

$$\begin{aligned}
\mathbf{0} =& \left[\sum_{t=1}^{T} \sum_{k=1}^{N} p(k|\mathbf{x}_t, \Theta_0) \cdot (\mathsf{M}_k)^T \cdot (\boldsymbol{\Sigma}_k^0)^{-1} \cdot \mathsf{M}_k\right] \cdot \boldsymbol{\alpha}^{\mathrm{ML}} \\
&- \sum_{t=1}^{T} \sum_{k=1}^{N} p(k|\mathbf{x}_t, \Theta_0) \cdot (\mathsf{M}_k)^T \cdot (\boldsymbol{\Sigma}_k^0)^{-1} \cdot \left(\mathbf{x}_t - \boldsymbol{\mu}_k^0\right).
\end{aligned} \tag{2.115}$$

The matrix M_k and the covariance matrix $\boldsymbol{\Sigma}_k^0$ are independent from the observations $\mathbf{x}_t$. According to equation (2.43), the sum over all observed feature vectors weighted by the posterior probability is equal to the ML estimate $\boldsymbol{\mu}_k^{\mathrm{ML}}$ multiplied by the soft number of assigned feature vectors n_k. This allows the following modification:

$$\mathbf{0} = \left[\sum_{k=1}^{N} n_k \cdot (\mathsf{M}_k)^T \cdot (\Sigma_k^0)^{-1} \cdot \mathsf{M}_k \right] \cdot \boldsymbol{\alpha}^{\mathrm{ML}} - \sum_{k=1}^{N} n_k \cdot (\mathsf{M}_k)^T \cdot (\Sigma_k^0)^{-1} \cdot (\boldsymbol{\mu}_k^{\mathrm{ML}} - \boldsymbol{\mu}_k^0) \,. \tag{2.116}$$

The solution of this set of equations delivers the optimal weigths $\boldsymbol{\alpha}^{\mathrm{ML}}$. The adapted mean vectors $\boldsymbol{\mu}_k^{\mathrm{EV}}$ result from the weighted sum of all eigenvoices in (2.101) given the optimal weights $\boldsymbol{\alpha}^{\mathrm{ML}}$.

In contrast to the MAP adaptation in Sect. 2.6.3 this approach is characterized by fast convergence even on limited training data due to prior knowledge about speaker variability. Only a few utterances are required. However, this also limits the adaptation accuracy for long-term adaptation [Botterweck, 2001]. Thus, EV adaptation is considered as a promising candidate for short-term adaptation, especially for the use case of this book.

2.6.6 Extended Maximum A Posteriori

The preceding sections presented some techniques for long-term and short-term adaptation. Now the question emerges how to handle the transition between limited and extensive training data. One way is to use an experimentally determined threshold. Hard decisions implicate the need for a reasonable choice of the associated thresholds and bear the risk to be not optimal.

The *Extended Maximum A Posterior* (EMAP) speaker adaptation presented by Stern and Lasry [1987] follows the strategy to integrate both aspects and offers an optimal transition. EMAP differs from the MAP adaptation in Sect. 2.6.3 since the parameters are regarded as random variables with statistical dependencies across all Gaussian densities [Stern and Lasry, 1987]. EMAP jointly adapts all Gaussian densities whereas MAP adaptation individually modifies each distribution.

The prior knowledge how the Gaussian densities of a codebook depend on the adaptation of the remaining density functions is employed to efficiently adjust even those which are not or not sufficiently trained [Jon et al., 2001] as shown in (2.125) further below. This behavior is similar to the EV technique and is important when adaptation relies only on limited data such as short utterances.

As soon as more utterances are accumulated, e.g. with the help of a robust speaker identification, each Gaussian density can be modified individually. Thus, in the extreme case of large data sets EMAP behaves like the MAP algorithm as shown later in (2.124).

First, some vectors and matrices are introduced so that EMAP adaptation can be described by the supervector notation introduced in the preceding section. All mean vectors $\boldsymbol{\mu}_k$ are stacked into a supervector

$$\check{\boldsymbol{\mu}} = \begin{pmatrix} \boldsymbol{\mu}_1 \\ \vdots \\ \boldsymbol{\mu}_k \\ \vdots \\ \boldsymbol{\mu}_N \end{pmatrix} \tag{2.117}$$

whereas the covariance matrix

$$\check{\boldsymbol{\Sigma}}^0 = \operatorname{diag}\left(\boldsymbol{\Sigma}_1^0 \ldots, \boldsymbol{\Sigma}_N^0\right) \tag{2.118}$$

represents a block matrix whose main diagonal consists of the covariances matrices $\boldsymbol{\Sigma}_k^0$ of the standard codebook. The covariance matrix

$$\check{\boldsymbol{\Sigma}}^{\mathrm{EMAP}} = \mathrm{E_i}\{(\check{\boldsymbol{\mu}}^i - \check{\boldsymbol{\mu}}^0) \cdot (\check{\boldsymbol{\mu}}^i - \check{\boldsymbol{\mu}}^0)^T\} \tag{2.119}$$

denotes a full covariance matrix representing prior knowledge about the dependencies between all Gaussian densities in analogy to (2.97). The estimation of this matrix is done in an off-line training.

At run-time all ML estimates given by (2.81) are stacked into a long concatenated vector $\check{\boldsymbol{\mu}}^{\mathrm{ML}}$. The number of softly assigned feature vectors

$$n_k = \sum_{t=1}^{T} p(k|\mathbf{x}_t, \Theta_0) \tag{2.120}$$

build the main diagonal of the matrix

$$\check{\mathrm{C}} = \operatorname{diag}\left(n_1, \ldots, n_N\right). \tag{2.121}$$

Stern and Lasry [1987] describe the EMAP approach in detail. Therefore the derivation is omitted in this book and only the result is given. The adapted mean supervector $\check{\boldsymbol{\mu}}^{\mathrm{EMAP}}$ can be calculated by

$$\check{\boldsymbol{\mu}}^{\mathrm{EMAP}} = \check{\boldsymbol{\mu}}^0 + \check{\boldsymbol{\Sigma}}^{\mathrm{EMAP}} \cdot (\check{\boldsymbol{\Sigma}}^0 + \check{\mathrm{C}} \cdot \check{\boldsymbol{\Sigma}}^{\mathrm{EMAP}})^{-1} \cdot \check{\mathrm{C}} \cdot (\check{\boldsymbol{\mu}}^{\mathrm{ML}} - \check{\boldsymbol{\mu}}^0) \tag{2.122}$$

as found by Jon et al. [2001], for example. Finally, the two extreme cases for $n_k \to 0, \forall k$ and $n_k \to \infty, \forall k$ should be discussed:

- When the number of the softly assigned feature vectors increases $n_k \to \infty$, the term $(\check{\boldsymbol{\Sigma}}^0 + \check{\mathrm{C}} \cdot \check{\boldsymbol{\Sigma}}^{\mathrm{EMAP}})^{-1}$ is dominated by $\check{\mathrm{C}} \cdot \check{\boldsymbol{\Sigma}}^{\mathrm{EMAP}}$ so that the following approximation may be employed:

$$\check{\boldsymbol{\mu}}^{\mathrm{EMAP}} \approx \check{\boldsymbol{\mu}}^0 + \check{\boldsymbol{\Sigma}}^{\mathrm{EMAP}} \cdot (\check{\mathrm{C}} \cdot \check{\boldsymbol{\Sigma}}^{\mathrm{EMAP}})^{-1} \cdot \check{\mathrm{C}} \cdot (\check{\boldsymbol{\mu}}^{\mathrm{ML}} - \check{\boldsymbol{\mu}}^0) \tag{2.123}$$

$$\boldsymbol{\mu}_k^{\mathrm{EMAP}} \approx \boldsymbol{\mu}_k^{\mathrm{ML}}, \ \forall k. \tag{2.124}$$

  Therefore, the MAP and EMAP estimates converge if a large data set can be used for adaptation.

- In the case of very limited data $n_k \to 0,\ \forall k$ the influence of $\check{\mathbf{C}} \cdot \check{\boldsymbol{\Sigma}}^{\text{EMAP}}$ decreases which allows the approximation

$$\check{\boldsymbol{\mu}}^{\text{EMAP}} \approx \check{\boldsymbol{\mu}}^{0} + \check{\boldsymbol{\Sigma}}^{\text{EMAP}} \cdot (\check{\boldsymbol{\Sigma}}^{0})^{-1} \cdot \check{\mathbf{C}} \cdot (\check{\boldsymbol{\mu}}^{\text{ML}} - \check{\boldsymbol{\mu}}^{0}). \tag{2.125}$$

  The matrix $\check{\boldsymbol{\Sigma}}^{\text{EMAP}} \cdot (\check{\boldsymbol{\Sigma}}^{0})^{-1}$ is independent from the adaptation data and represents prior knowledge. Even limited adaptation data can be handled appropriately. The intention is therefore similar to the EV approach discussed in Sect. 2.6.5.

One disadvantage of the EMAP technique is the high computational complexity. Especially, the matrix inversion in (2.122) is extremely demanding because the dimension of $\check{\boldsymbol{\Sigma}}^{0} + \check{\mathbf{C}} \cdot \check{\boldsymbol{\Sigma}}^{\text{EMAP}}$ is the product of the number of Gaussian densities N and the dimension of the feature vectors. Unfortunately, the inversion contains the number of the softly assigned feature vectors n_k so that this inversion has to be done at run-time [Rozzi and Stern, 1991]. Jon et al. [2001] approximate the covariance matrix $\check{\boldsymbol{\Sigma}}^{\text{EMAP}}$ by some 25 eigenvectors. This approximation helps to significantly reduce the complexity of the matrix inversion.

2.7 Feature Vector Normalization and Enhancement

In this section the normalization and enhancement of feature vectors is introduced. Alternative approaches are described to deal with speaker variabilities. In contrast to speaker adaptation unwanted variations, e.g. caused by environmental effects or speakers [Song and Kim, 2005], are removed from the speech signal and are not integrated in the statistical modeling.

First, a motivation for feature normalization and enhancement as well as a brief classification of the existing techniques are given to demonstrate the change of the procedural method to handle speaker variabilities. Then a selection of approaches known from the literature is presented.

2.7.1 Motivation

The main aspect of speaker adaptation is to alleviate the mismatch between training and test situations. The source-filter theory in Sect. 2.1 gives reasons for a possible mismatch by speaker specific characteristics, channel impulse responses and background noises, for example.

Speaker adaptation looks for a better statistical representation of speech signals by integrating these discrepancies into the statistical model. Even though speaker adaptation is considered to produce normally the best results, it is more computationally demanding than speech normalization in the feature space [Buera et al., 2007].

Normalization and enhancement follow the idea to remove a mismatch between the test and training conditions, e.g. due to environmental disturbances or speaker dependencies, from the feature vectors. These fluctuations should not be visible for speech recognition.

Normalization and enhancement of feature vectors can be advantageous when users operate speech controlled devices under various environmental conditions and when speaker adaptation can adjust only one statistical model for each speaker. Therefore, techniques for feature vector enhancement are required which are able to handle different mismatch conditions [Song and Kim, 2005].

Noise reduction and cepstral mean normalization join this procedure because they limit the effects of background noises or room impulse responses on the feature vector space.

Vocal Tract Length Normalization (VTLN) may be applied in addition. Inter-speaker variability due to physiological differences of the vocal tract can be moderated by a speaker-specific frequency warping prior to cepstral feature extraction [Garau et al., 2005; Kim et al., 2004], e.g. by modifying the center frequencies of the mel filter bank [Garau et al., 2005; Häb-Umbach, 1999].

A variety of additional methods have been developed which perform a correction directly in the feature space. Feature normalization and enhancement can be divided into three categories [Buera et al., 2007]:

- **High-pass filtering** comprises cepstral mean normalization and relative spectral amplitude processing [Buera et al., 2007], for example.
- **Model-based techniques** presume a structural model to describe the environmental degradation and apply the inverse operation for the compensation at run-time [Buera et al., 2007].
- **Empirical compensation** apply statistical models which represent the relationship between the clean and disturbed speech feature vectors. They enable the correction or mapping of the observed disturbed feature vectors onto their corresponding clean or undisturbed representatives.

Two algorithms are presented in Sect. 2.7.2 and Sect. 2.7.3 which apply similar statistical models compared to the methods for speaker identification and speaker adaptation described in the preceding sections.

2.7.2 Stereo-based Piecewise Linear Compensation for Environments

The *Stereo-based Piecewise Linear Compensation for Environments* (SPLICE) technique is a well-known approach for feature vector based compensation [Buera et al., 2007; Droppo et al., 2001; Moreno, 1996]. It represents a non-linear feature correction as a front-end of a common speech recognizer [Droppo and Acero, 2005].

A statistical model is trained to learn the mapping of undisturbed feature vectors to disturbed observations. At run-time the inverse mapping is performed. The disturbed feature vectors $\mathbf{y}_t$ have to be replaced by optimal estimates, e.g. given by the *Minimum Mean Squared Error* (MMSE) criterion.

MMSE estimates minimize the Euclidean distance between the correct clean feature vectors and the resulting estimates on average. The optimal estimate $\mathbf{x}_t^{\mathrm{MMSE}}$ is given by the expectation of the corresponding conditional probability density function $p(\mathbf{x}_t|\mathbf{y}_t)$ [Droppo and Acero, 2005; Hänsler, 2001; Moreno, 1996] as shown below:

$$\mathbf{x}_t^{\mathrm{MMSE}} = \mathrm{E}_{\mathbf{x}|\mathbf{y}}\{\mathbf{x}_t\} \tag{2.126}$$

$$\mathbf{x}_t^{\mathrm{MMSE}} = \int_{\mathbf{x}_t} \mathbf{x}_t \cdot p(\mathbf{x}_t|\mathbf{y}_t)\,\mathrm{d}\mathbf{x}_t. \tag{2.127}$$

A latent variable may be introduced to represent multi-modal density functions. A more precise modeling can be achieved. The sum over all density functions

$$\mathbf{x}_t^{\mathrm{MMSE}} = \int_{\mathbf{x}_t} \mathbf{x}_t \cdot \sum_{k=1}^{N} p(\mathbf{x}_t, k|\mathbf{y}_t)\,\mathrm{d}\mathbf{x}_t \tag{2.128}$$

reduces the joint density function $p(\mathbf{x}_t, k|\mathbf{y}_t)$ to the conditional density function $p(\mathbf{x}_t|\mathbf{y}_t)$. A further simplification is achieved by using Bayes' theorem

$$\mathbf{x}_t^{\mathrm{MMSE}} = \int_{\mathbf{x}_t} \mathbf{x}_t \cdot \sum_{k=1}^{N} p(\mathbf{x}_t|k, \mathbf{y}_t) \cdot p(k|\mathbf{y}_t)\,\mathrm{d}\mathbf{x}_t. \tag{2.129}$$

Integral and sum can be interchanged

$$\mathbf{x}_t^{\mathrm{MMSE}} = \sum_{k=1}^{N} p(k|\mathbf{y}_t) \cdot \int_{\mathbf{x}_t} \mathbf{x}_t \cdot p(\mathbf{x}_t|k, \mathbf{y}_t)\,\mathrm{d}\mathbf{x}_t \tag{2.130}$$

$$= \sum_{k=1}^{N} p(k|\mathbf{y}_t) \cdot \mathrm{E}_{\mathbf{x}|\mathbf{y},\mathbf{k}}\{\mathbf{x}_t\} \tag{2.131}$$

due to the linearity of the expectation value.

Under the assumption that the undisturbed feature vector $\mathbf{x}_t$ can be represented by the observation $\mathbf{y}_t$ and an offset vector [Moreno, 1996], equation (2.131) can be realized by

$$\mathbf{x}_t^{\mathrm{MMSE}} = \mathbf{y}_t + \sum_{k=1}^{N} p(k|\mathbf{y}_t) \cdot \mathbf{r}_k. \tag{2.132}$$

The offset vectors $\mathbf{r}_k$ have to be trained on a stereo data set that comprises both the clean feature vectors $\mathbf{x}$ and the corresponding disturbed feature vectors $\mathbf{y}$ of the identical utterance.

The posterior probability $p(k|\mathbf{y}_t)$ determines the influence of each offset vector. An additional GMM is introduced

$$p(\mathbf{y}_t) = \sum_{k=1}^{N} w_k \cdot \mathcal{N}\left\{\mathbf{y}_t | \boldsymbol{\mu}_k, \Sigma_k\right\} \tag{2.133}$$

which represents the distribution of the observed feature vectors $\mathbf{y}_t$ as found by Buera et al. [2007], for example. The posterior probability $p(k|\mathbf{y}_t)$ can be calculated by combining (2.133) and (2.41).

This SPLICE technique only marks a basic system since a manifold of variations and extensions exist. Buera et al. [2005] should be mentioned because they investigate similar techniques in the context of speaker verification and speaker identification.

2.7.3 Eigen-Environment

The SPLICE algorithm is built on a mixture of conditional distributions $p(\mathbf{x}_t|\mathbf{y}_t, k)$ each contributing to the overall correction. The goal is to deliver an optimal estimate $\mathbf{x}_t^{\mathrm{MMSE}}$ of the clean feature vector.

Song and Kim [2005] describe an eigen-environment approach that is a combination of the SPLICE method and an eigenvector technique similar to the EV approach. The main issue is the computation or approximation of this offset vector $\mathbf{r}_k$ by a linear combination of eigenvectors. The basic directions of the offset vectors $\mathbf{r}_k$ and the average offset vector $\mathbf{e}_{k,0}^{\mathrm{Env}}$ have to be extracted in a training similar to the EV adaptation. Only the first L eigenvectors $\mathbf{e}_{k,l}^{\mathrm{Env}}$ associated with the highest eigenvalues are employed.

At run-time a set of scalar weighting factors α_l has to be determined to obtain an optimal approximation of the offset vector

$$\mathbf{r}_k = \mathbf{e}_{k,0}^{\mathrm{Env}} + \sum_{l=1}^{L} \alpha_l \cdot \mathbf{e}_{k,l}^{\mathrm{Env}}. \tag{2.134}$$

Song and Kim [2005] estimate the parameters α_l with the help of the EM algorithm by maximizing the auxiliary function Q_{ML} in a similar way as the EV adaptation in Sect. 2.6.5. The estimate of the undisturbed feature vector can then be calculated by (2.132).

As discussed before, an auxiliary GMM is required to represent the distribution of the disturbed feature vectors. It enables the application of (2.41) to compute the posterior probability $p(k|\mathbf{y}_t)$ which controls the influence of each offset vector.

2.8 Summary

In this chapter the basic components of a complete system comprising speaker change detection, speaker identification, speech recognition and speaker adaptation have been described.

The discussion started with the human speech production and introduced the source-filter model. The microphone signal is explained by excitation, vocal tract, channel characteristics and additional noise. Speaker specific excitation and vocal tract as well as acquired speaker habits can contribute to speaker variability [Campbell, 1997; O'Shaughnessy, 2000].

The front-end compensates for noise degradations and channel distortions and provides a compact representation of the spectral properties suitable for speech recognition and speaker identification.

Several statistical models were introduced to handle speech and speaker variability for speaker change detection, speaker identification and speech recognition:

BIC as a well-known method for speaker change detection uses multivariate Gaussian distributions for the statistical modeling of hypothetical speaker turns. Speaker identification extends this model by applying a mixture of Gaussian distributions to appropriately capture speaker characteristics such as the vocal tract. Speech recognition is usually not intended to model particular speaker characteristics but has to provide a transcription of speech utterances. Dynamic speech properties have to be taken into consideration. For this purpose HMMs combine GMMs with an underlying Markov process to improve the modeling of speech.

Finally, speaker adaptation provides several strategies to modify GMMs and HMMs and targets to reduce mismatch situations between training and test. The main difference of the techniques described in this chapter can be seen in the degrees of freedom which have to be balanced with the amount of available training data. Short-term and long-term adaptation have been explained in detail.

Normalization or enhancement provides a different approach to deal with speaker variability. Instead of learning mismatch conditions, they are removed from the input signal. Even though speaker adaptation generally yields better results [Buera et al., 2007], applications are possible which combine those strategies. In fact, the combination of cepstral mean subtraction and speaker adaptation already realizes a simple implementation.

More complex systems, e.g. speaker tracking or speaker specific speech recognition, can be constructed based on these components as described in the following chapters.

3

Combining Self-Learning Speaker Identification and Speech Recognition

Chapter 2 introduced the fundamentals about speaker change detection, speaker identification, speech recognition and speaker adaptation. The basic strategies relevant for the problem of this book were explained.

In this chapter a survey of the literature concerning the intersection of speaker change detection, speaker identification, speaker adaptation and speech recognition is given. The approaches are classified into three groups. They mainly differ in the complexity of the speaker specific models and the methods of involving speech and speaker characteristics.

A first conclusion with respect to the cited literature is drawn in the last section. The advantages and drawbacks of the approaches are discussed and the target system of this book is sketched.

3.1 Audio Signal Segmentation

A frequently investigated scenario can be seen in the broad class of audio segmentation. A continuous stream of audio data has to be divided into homogeneous parts [Chen and Gopalakrishnan, 1998; Hain et al., 1998; Lu and Zhang, 2002; Meinedo and Neto, 2003]. The criteria for homogeneity may be the acoustic environment, channel properties, speaker identity or speaker gender [Chen and Gopalakrishnan, 1998; Meinedo and Neto, 2003]. This procedure is known as segmentation, end-pointing or divisive segmentation.

The problem of speaker change detection is to find the beginning and end of an utterance for each speaker. The distinction of speech and non-speech parts can be an additional task [Meinedo and Neto, 2003]. In other words the question to be solved is which person has spoken when [Johnson, 1999].

Speaker change detection has found its way into a broad range of applications. It can be used for broadcast news segmentation to track the anchor speakers [Johnson, 1999], for example. Further applications are automatic transcription systems for conferences or conversations [Lu and Zhang, 2002]. Automatic transcription systems aim at journalizing conversations and at

T. Herbig, F. Gerl, and W. Minker: Self-Learning Speaker Identification, SCT, pp. 59–70.
springerlink.com

assigning the recognized utterances to the associated persons. Video content analysis and audio/video retrieval can be enhanced by unsupervised speaker change detection [Lu and Zhang, 2002]. The ability to track the speaker enables more advanced speaker adaptation techniques and reduces the error rate of unsupervised transcription systems [Kwon and Narayanan, 2002].

Speaker change detection has to handle data streams typically without prior information about the beginning and end of a speaker's utterance, speaker identity or the number of speakers [Lu and Zhang, 2002]. A high number of speaker changes has to be expected a priori for applications such as meeting indexing. Real-time conditions can be a further challenge [Lu and Zhang, 2002].

Divisive segmentation splits the data stream into segments that contain only one speaker, for example. The given data segment is tested for all theoretical speaker turns as illustrated in Fig. 3.1(a). This can be realized by fixed-length segmentation or by a variable window length [Kwon and Narayanan, 2002].

Fixed-length segmentation partitions the data stream into small segments of predefined length by applying overlapping or non-overlapping window functions. For example, Cheng and Wang [2003] use a first segmentation with a window length of 12 sec and a refined window of $2 - 3$ sec. Each boundary is checked for a speaker change. If no change is found both blocks are merged and the next boundary is tested. Small segments, however, may complicate robust end-pointing, e.g. in the case of small speech pauses [Kwon and Narayanan, 2002].

Alternatively, the window length can be dynamically enlarged depending on the result of the speaker change detection [Ajmera et al., 2004]. This enables the system to estimate more parameters. However, both the computational load and time delays limit the application of complex statistical models and extensive training algorithms for real-time applications [Lu and Zhang, 2002].

This problem can be extended to speaker tracking. In this case an automatic speaker indexing has to be performed to retrieve all occurrences of one speaker in a data stream [Meinedo and Neto, 2003]. Hence, *agglomerative clustering* can be regarded as the continuation of divisive clustering. The clustering can be solved by using a two step procedure. First, the data stream is divided into homogeneous blocks. Then, those blocks which are assumed to correspond to the same speaker are combined. For example, the same or similar algorithms can be applied to the result of the speaker change detection to test the hypothesis that non-adjacent segments originate from one speaker. This strategy allows building larger clusters of utterances comprising only one speaker. Thus, more robust and complex speaker specific models can be trained [Mori and Nakagawa, 2001]. An example is given in Fig. 3.1(b).

Chen and Gopalakrishnan [1998]; Hain et al. [1998]; Johnson [1999]; Meinedo and Neto [2003]; Tritschler and Gopinath [1999]; Yella et al. [2010];

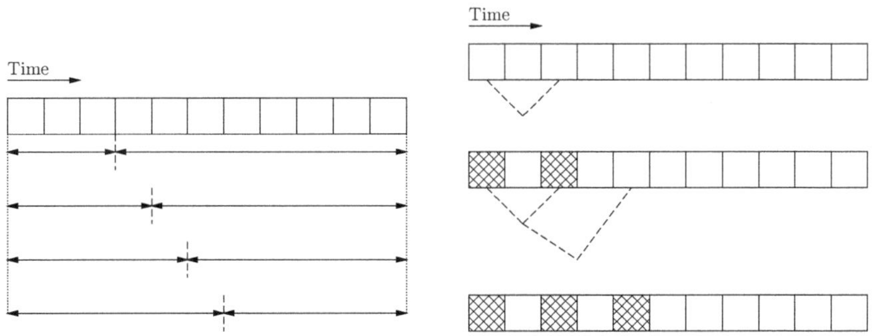

(a) Divisive segmentation. (b) Agglomerative clustering.

Fig. 3.1 Comparison of divisive segmentation (a) and agglomerative clustering (b). Divisive segmentation splits the continuous data stream into data blocks of predefined length which are iteratively tested for speaker turns. Based on this result the same or other algorithms are applied for agglomerative clustering to merge non-adjacent segments if they are assumed to originate from the same speaker.

Zhou and Hansen [2000] provide further detailed descriptions and several implementations.

Segmentation and clustering can be structured into three groups [Chen and Gopalakrishnan, 1998; Cheng and Wang, 2003; Kwon and Narayanan, 2002]:

- **Metric-based** techniques compute the distances of adjacent intervals of the considered audio signal [Kwon and Narayanan, 2002]. Maxima of the distance measure indicate possible speaker changes and a threshold detection leads to the acceptance or rejection of the assumed speaker turns. Metric-based techniques are characterized by low computational cost but require experimentally determined thresholds [Cheng and Wang, 2003]. Furthermore, they usually span only short time intervals and thus may not be suitable for robust distance measures [Cheng and Wang, 2003]. For example, the Euclidean distance or an information theoretic distance measure can be applied as found by Kwon and Narayanan [2002]; Siegler et al. [1997]; Zochová and Radová [2005].
- **Model-based** algorithms such as BIC estimate statistical models for each segment and perform a hypothesis test. These models are locally trained on the segment under investigation. A low model complexity is required because of limited data. A detailed description of the BIC was given in Sect. 2.3 or can be found by Ajmera et al. [2004]; Chen and Gopalakrishnan [1998]; Tritschler and Gopinath [1999]; Zhou and Hansen [2000].

- **Decoder-guided** techniques extend this model-based procedure by using pre-trained statistical models. When speaker models are trained off-line, more sophisticated models of higher complexity and more extensive training algorithms can be applied. This approach circumvents the restriction of locally optimized statistical modeling at the cost of an enrollment. At run-time the Viterbi algorithm can be applied to determine the most likely sequence of speaker occurrences in a data stream as found by Hain et al. [1998]; Wilcox et al. [1994]. For example, the audio stream can be processed by a speech recognizer. The transcription delivers the candidates for speaker turns, e.g. speech pauses, which can be employed in the subsequent speaker segmentation [Zochová and Radová, 2005].

Nishida and Kawahara [2005] present a multi-stage algorithm to cluster utterances based on speaker change detection or statistical speaker model selection. In the first implementation the variance BIC is applied for divisive and agglomerative clustering. The second proposed framework operates either on discrete or continuous speaker models depending on the amount of training data. Robustness is gained by a simple discrete model during the initialization phase. The system switches in an unsupervised manner to GMM modeling when a robust parameter estimation becomes feasible. The transition is realized by the Gaussian mixture size selection that is set up on the BIC framework. Identification and verification methods enable the system to merge the speaker clusters.

The result of a robust audio signal segmentation can be employed to realize speaker specific speech recognition. As long as the current speaker does not change the speech recognizer is able to continuously adapt the speech model to the speaker's characteristics.

Zhang et al. [2000] investigate a two-stage procedure of speaker change detection and speech recognition. The first stage checks an audio signal for speaker turns and the second stage decodes the speech signal with a refined speaker adapted HMM. During the first stage one speaker independent HMM and up to two speaker specific HMMs are evaluated in parallel. The ML criterion is used to decide whether a new speaker has to be added to the recognition system and whether a speaker change has to be assumed. If a new speaker is indicated, a new HMM is initialized by adapting the speaker independent HMM to the current utterance. The updated speaker specific HMM is then used for re-recognition of the utterance and delivers the final transcription result. In a second experiment the HMM recognition system is extended by speaker specific GMMs and a further UBM to decrease the computational effort. The UBM is realized by a GMM comprising 64 Gaussian distributions. Speaker change detection is implemented comparably to the detection of unknown speakers in Sect. 2.4.3. The most likely GMM is selected according to the ML criterion and the associated HMM is used for speech decoding.

3.2 Multi-Stage Speaker Identification and Speech Recognition

Many algorithms in audio signal segmentation such as BIC use simple statistical models to detect speaker changes even on few speech data. Some of these algorithms were described in the preceding section.

Speaker identification usually relies on more complex models such as GMMs which require a sufficient amount of training and test data. Some publications employ the combination of both strategies as will be shown in this section.

Figure 3.2(a) displays a possible realization of such an approach. The feature extraction transforms the audio signal into a vector representation comprising discriminatory features of the speech signal. VAD separates speech and speech pauses so that only the speech signals have to be processed. Speaker change detection clusters the feature vectors of an utterance and assures that only one speaker is present. Speaker identification determines the speaker identity and selects the corresponding speaker model. The subsequent speaker adaptation applies common adaptation techniques to better capture speaker characteristics. If speaker identification is combined with speech recognition, the knowledge about the speaker identity also enables speaker specific speech recognition.

Geiger et al. [2010] present a GMM based system for open-set on-line speaker diarization. In the first stage audio segmentation is performed frame by frame. A threshold criterion and a rule-based framework are applied to determine the start and end points of speech and garbage words based on the energy of the audio signal. MFCCs are extracted to be used for several classification steps, e.g. speech or non-speech as well as male or female. Three GMMs are trained off-line - male, female and garbage. They are employed to classify each segment and to identify known speakers. When one of the gender models yields the highest likelihood score, a new speaker is assumed and a new model is initialized. MAP adaptation is used to initialize additional GMMs and to continuously adapt the speaker models.

Wu et al. [2003] investigate speaker segmentation for real-time applications such as broadcast news processing. A two-stage approach is presented comprising pre-segmentation and refinement. Pre-segmentation is realized by a UBM which categorizes speech into three categories - reliable speaker-related, doubtful speaker-related and unreliable speaker-related. Based on reliable frames the dissimilarity between two segments is computed and potential speaker changes are identified. In the case of an assumed speaker turn this decision is verified by adapted GMMs. If no speaker change is found, incremental speaker adaptation is applied to obtain a more precise speaker model. Otherwise, the existing model is substituted by a new model initialized from UBM.

Kwon and Narayanan [2005] address the issue of unsupervised speaker indexing by developing a complete system combining speaker change detection, speaker identification and speaker adaptation. Their algorithm presumes no prior knowledge about speaker changes or the number of speakers. VAD filters the speech signal and rejects all non-speech parts such as background noise. Only the speech signal is further processed by the speaker change detection which applies model-based segmentation algorithms as introduced in Sect. 3.1 to detect possible speaker turns. As soon as a different speaker is assumed, the data between the last two speaker turns is clustered and a pool of speaker models is incrementally built up and refined. The problem of model initialization is solved by a set of reference speakers. The acoustic similarity between particular speakers is exploited by a Monte Carlo algorithm to construct an initial speaker model that can be further adapted to the new speaker. This approach is called sample speaker model and is compared to the UBM and Universal Gender Model (UGM) initialization method for speaker adaptation. The final speaker adaptation is implemented by MAP adaptation.

Further multi-stage procedures combining segmentation and clustering are described by Hain et al. [1998], for example.

As soon as a confident guess concerning the speaker's identity is available speaker specific speech recognition becomes feasible as found by Mori and Nakagawa [2001]. Nishida and Kawahara [2004] present an approach that combines automatic speech recognition with a preceding speaker indexing to obtain a higher recognition accuracy. Speaker identification controls speech recognition so that the associated speaker specific speech models can be used to enhance speech recognition.

These speaker indexing algorithms consist of a step-by-step processing. VAD makes discrete decisions in speech and non-speech segments and the speaker change detection controls the clustering and adaptation algorithm. Therefore no confidence measure about previous decisions is available in the subsequent processing. This loss of information can be circumvented when a system is based on soft decisions and when the final discrete decisions, e.g. speaker identity, are delayed until speaker adaptation.

Schmalenstroeer and Häb-Umbach [2007] describe an approach for speaker tracking. Speech segmentation, speaker change detection, speaker identification and spatial localization by a beamformer[1] are considered simultaneously. All cues for speaker changes, speaker identity and localization can be kept as soft quantities. This prevents a loss of information caused by threshold detections. Each speaker represents a state of an HMM and the associated GMM contains the speaker characteristics. The transitions of the Markov chain can be controlled by a speaker change or tracking algorithm. The most likely sequence of speaker identities is determined by the Viterbi algorithm to achieve a more robust speaker tracking. This approach provides a method of a complete speaker tracking system.

[1] Details on beamforming can be found by Krim and Viberg [1996]; Veen and Buckley [1988], for example.

3.3 Phoneme Based Speaker Identification

The goal of the algorithms in the preceding sections was to control speech recognition by estimating the speaker identity. The opposite problem is also known from literature. Speech recognition can enhance speaker identification since some groups of phonemes differ in their discriminative information.

Figure 3.2(b) displays a simplified block diagram of phoneme based speaker identification. It is pointed out that speaker identification and speech recognition are interchanged compared to Fig. 3.2(a).

The configuration of the articulatory organs widely varies during speech production as briefly described in Sect. 2.1. Hence, speaker specific voice characteristics are not accurately modeled by averaged features. Acoustic discrimination can improve the performance of speaker identification [Rodríguez-Liñares and García-Mateo, 1997].

Eatock and Mason [1994]; Harrag et al. [2005]; Pelecanos et al. [1999]; Thompson and Mason [1993] agree that several phoneme groups exhibit a higher degree of speaker discriminative information. Thus, a phoneme recognition before speaker identification can enhance the speaker identification rate. These publications conclude that vowels and nasals contain the highest speaker specific variability whereas fricatives and plosives have the lowest impact on speaker identification.

This observation motivated Kinnunen [2002] to construct a speaker dependent filter bank. The key issue is to emphasize specific frequency bands which are characteristic for a considered group of phonemes to account for speaker characteristics. The goal is an improved representation of the speaker discriminative phonemes to increase the speaker identification rate.

Gutman and Bistritz [2002] apply phoneme-based GMMs to increase the identification accuracy of GMMs. A stronger correlation of phonemes and Gaussian mixtures is targeted. For each speaker one GMM is extensively trained by standard training algorithms without using a phoneme classification. Phoneme segmentation is then used to cluster the speaker's training data. The phonetic transcription of the TIMIT database and the Viterbi algorithm are used for the classification. This scenario is similar to a text-prompted speaker verification. Finally, a set of phoneme dependent GMMs is trained for all speakers by adapting the phoneme independent GMM to each phoneme cluster. A speaker independent background model is equally constructed. For testing, the likelihoods of all speaker specific GMMs are computed given the phoneme segmentation. They are normalized by the corresponding background model. The ratio of both likelihoods is evaluated to accept or reject the identity claim of the current speaker. In their tests, a lower error rate was achieved for speaker verification based on phoneme-adapted GMMs compared to a standard GMM approach.

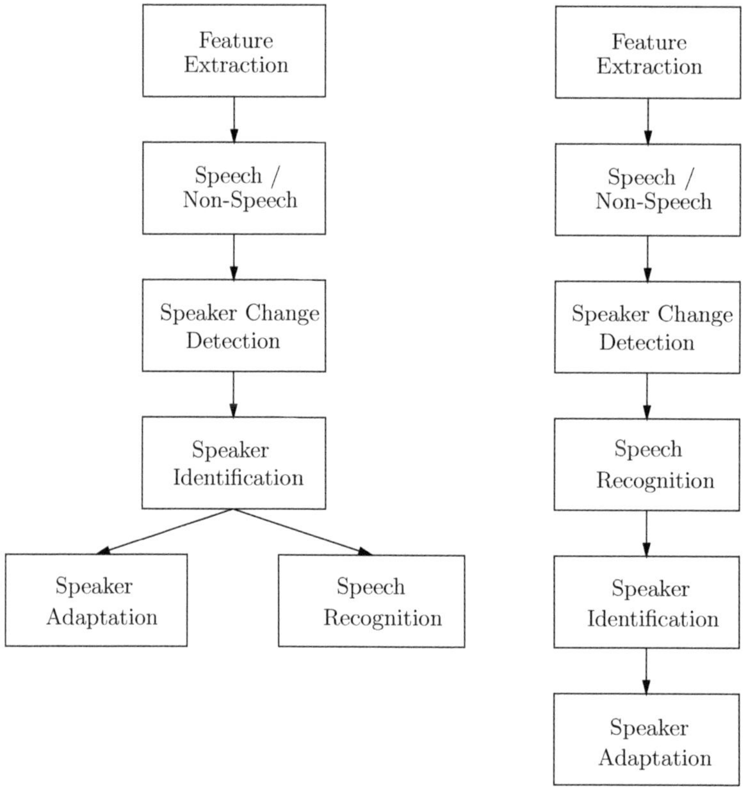

(a) Exemplary block diagram of a multi-stage speaker identification including speech recognition and speaker adaptation.

(b) Exemplary block diagram of a phoneme based speaker identification with integrated speaker adaptation.

Fig. 3.2 Comparison of two examples for multi-stage speaker identification and phoneme based speaker identification. In both cases the result of the speaker change detection triggers the recognition. Either the most probable speaker is identified to enable an enhanced speech recognition or the spoken phrases is divided into phoneme groups by a speech recognizer to enhance speaker identification. In addition, speaker adaptation can be continuously applied to the speech and speaker models.

Gauvain et al. [1995] present a statistical modeling approach for text-dependent and text-independent speaker verification. Phone-based HMMs are applied to obtain a better statistical representation of speaker characteristics. Each phone is modeled by a speaker-specific left-to-right HMM comprising three states. 75 utterances are used for training which is realized by MAP adaptation of an initial model comprising 35 speaker-independent context-free phone models and 32 Gaussian distributions. For testing all speaker models are run in parallel. For each speaker the phone-based likelihood is computed by the Viterbi algorithm and the ML criterion is used to hypothesize the speaker identity. For the text-independent verification task local phonotactic constraints are imposed to limit the search space. For text-dependent speaker verification the limitation is given by the concatenated phone sequence. Both high-quality and telephone speech are examined.

Kimball et al. [1997] describe a two-stage approach for text-dependent speaker verification. Voice signatures are investigated for electronic documents. A common speech recognizer categorizes the speech input into several phonetic classes to train more discriminative statistical models for speaker verification. This representation is also known as the broad phonetic category model. The second approach applies supervised speaker adaptation to speaker specific HMMs. The evaluation of the HMMs associated with the given text is combined with cohort speaker normalization to solve the verification task.

Olsen [1997] also uses a two-stage approach for speaker identification. The initial stage consists of an HMM for phone alignment whereas an RBF is applied in the second stage for speaker identification given the phonetic information.

Genoud et al. [1999] present a combination of speech recognition and speaker identification with the help of a neural network. This approach offers the opportunity of a mutual benefit because speech and speaker information are modeled and evaluated simultaneously. The discriminative MLP receives MFCC features at its input layer and is trained for both tasks. It comprises two sets of output. Besides the speaker independent phoneme output, the posterior probability of each phoneme is examined for several target speakers. In their experiments 12 registered speakers are investigated in a closed-set scenario. The generalization towards open-set systems including unsupervised adaptation at the risk of error propagation is left for future research.

Park and Hazen [2002] also suggest to use a two-stage approach by combining common GMM based speaker identification and phoneme based speaker identification. The latter relies on the phonetically structured statistical models introduced by Faltlhauser and Ruske [2001] or phonetic class GMMs. The main issue is to use granular GMMs which resolve the phonetic content more precisely compared to globally trained GMMs. In addition, the speaker adaptive scoring allows the system to learn the speaker specific phoneme pronunciation when sufficient speech material can be collected for a particular speaker. The proposed identification framework hypothesizes the temporal phoneme alignment by a speaker independent speech recognizer given a spoken phrase. In parallel, GMMs are applied to pre-select a list of the best matching speaker models which have to be processed by the second stage. There, the phonetically refined speaker models re-score the subset of selected speakers and lead to the final speaker identification. In summary, the speech recognition result controls the selection of the speaker models and exploits more information about the speech utterance than globally trained GMMs.

Rodríguez-Liñares and García-Mateo [1997] combine speech classification, speech segmentation and speaker identification by using an HMM system.

Speech signals are divided into voiced and unvoiced speech as well as the associated transitions. In addition, speech pauses representing background noises are modeled. These categories are modeled by four HMMs. At run time the phonetic classification is implemented by the Viterbi algorithm. The probabilities of all phonetic classes are accumulated and combined either by equal weighting or by applying selective factors to investigate only the effect of single classes. The speaker with the highest score is selected. In addition, the speech segmentation is obtained as a by-product. In summary, speaker specific information is mainly contained in voiced parts of an utterance. Thus, the speaker identification rate on purely voiced parts is significantly higher than for unvoiced speech.

Matsui and Furui [1993] consider speech recognition and speaker verification in a unified frame work. The task is to verify the identity claim of the actual user by his voice characteristics and the spoken utterance that is prompted by the system. Matsui and Furui represent speakers by speaker specific phoneme models which are obtained by adaptation during an enrollment. At run-time the convex combination of both likelihoods of a phoneme dependent and a phoneme independent speaker HMM enables the acceptance or rejection of the claimed speaker identity. The result is a combined statistical model for text-dependent speech recognition and speaker verification.

Reynolds and Carlson [1995] compare two approaches for text-dependent speaker verification. The user is asked to speak 4 randomized combination locks prompted by the system. It has to be decided either to accept or reject the claimed identity. The first technique investigates a parallel computation of text-independent speaker verification and speaker independent speech recognition. The task is to verify both the speaker's identity and the uttered phrase. A decision logic combines the single scores of the decoupled identifiers. The second method consists of parallel speaker specific speech recognizers which perform both tasks. Each speaker undergoes an enrollment phase of approximately 6 min and uses a limited vocabulary. Cohort speakers serve as alternative models and are used for score normalization.

Nakagawa et al. [2006] propose a combined technique comprising two statistical models. The first model is a common GMM known from the speaker identification literature and the second one is an HMM well known from speech recognition. Both models have shown their strengths in representing speaker and speech information. The second model is motivated by the fact that temporal transitions between different phonemes are not modeled by GMMs. They can be captured by approaches coming from the speech recognition research. Their statistical model incorporates the knowledge about speech characteristics by adapting a syllable-based speaker independent speech model to particular speakers. At run-time both systems run in parallel and a convex combination of the independent scores integrates both the text-independent speaker information and speech related knowledge.

3.4 First Conclusion

All approaches touched on in the preceding sections have advantages and drawbacks with respect to the use case of this book.

The algorithms belonging to audio signal segmentation usually offer the advantage of statistical models of low complexity which only require a small number of parameters to be estimated. This makes short-term speaker adaptation feasible but also limits the capability to represent the individual properties of a speaker which could be modeled using long-term speaker adaptation. Another strong point is the opportunity to omit the training to increase the usability of speech controlled devices. However, most of these algorithms do not easily support to build-up a more complex system. In general, it seems not possible to construct highly complex statistical models for speech and speaker characteristics in an unsupervised manner without sophisticated initial models [Kwon and Narayanan, 2005]. Furthermore, many of these algorithms bear the risk to be vulnerable against time-variant adverse environments which is the case for in-car applications. Time latency is a further critical issue for the use case of this book.

Multi-stage speaker identification overcomes the limitation of simple statistical models. For example, prior knowledge is provided in form of an initial model which can be adapted to particular speakers. Some techniques tackle the problem of speaker change detection and speaker identification simultaneously and thus motivate a unified framework. However, a training phase for each new speaker is usually obligatory. Furthermore, some implementations apply a strict multi-step procedure with separate modules. Each module receives only the discrete results from the preceding stages. The final result is often obtained by the ML criterion and does not provide confidence measures. A further drawback is the multiple computation of speech and speaker information even though similar statistical models and similar or even identical input are applied.

Phoneme based speaker identification introduces a new aspect into speaker identification. The identification of the actually spoken phonemes and the knowledge that speaker identification is directly influenced by certain phoneme groups allow constructing more precise speaker models at the expense of a more complex training phase. Once again the separate modeling of speech and speakers and the application of multi-stage procedures may be of disadvantage. This information flow is usually unidirectional. Discrete decisions and the lack of confidence measures may be additional drawbacks.

The goal of this book is to present a solution which integrates speech and speaker information into one statistical model. Furthermore, speech and speaker shall be recognized simultaneously to allow real-time processing on embedded devices. The intended functionality of the desired solution is sketched in Fig. 3.3.

The critical issues are the self-organization of the complete system, detecting unknown speakers and avoiding a supervised training phase. The system

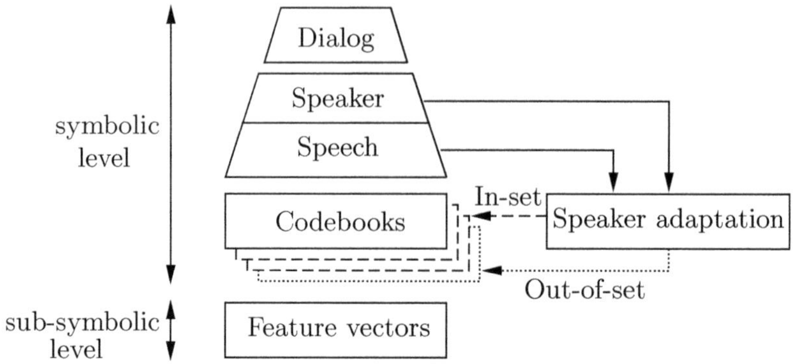

Fig. 3.3 Draft of the target system's functionality. The speech controlled system shall comprise speech recognition, speaker identification and speaker adaptation. Interactions with a speech dialog should be possible. In the sub-symbolic level feature vectors are extracted to be used in the symbolic levels. Several profiles or codebooks are used for speech recognition. A speaker independent codebook (solid line) acts as template for further speaker specific codebooks (dashed line). The results of speaker identification and speech recognition trigger speaker adaptation either to adjust the codebooks of known speakers (in-set) or to initialize new codebooks (dotted line) in the case of a new speaker (out-of-set). Significantly higher speech recognition rates are expected compared to speaker independent speech recognition.

has to handle speaker models at highly different training levels and has to provide a unified adaptation technique which permits both short-term and long-term adaptation. The system does not start from scratch but is supported by a robust prior statistical model and adapts to particular speakers. In the beginning, only a few adaptation parameters can be reliably estimated. This number may be increased depending on the training level. Speaker identification has to be realized on different time scales. The entire information is preserved by avoiding discrete decisions. Confidence measures in form of posteriori probabilities instead of likelihoods have to be established.

4

Combined Speaker Adaptation

The first component of the target system is a flexible yet robust speaker adaptation to provide speaker specific statistical models. In this chapter a balanced adaptation strategy is motivated. New speaker profiles can be initialized and existing profiles can be continuously adapted in an unsupervised way. The algorithm is suitable for short-term adaptation based on a few utterances and long-term adaptation when sufficient training data is available. A first experiment has been conducted to investigate the improvement of a speech recognizer's performance when speakers are optimally tracked so that the corresponding speaker models can be continuously adapted.

In the subsequent chapters, this adaptation technique will serve as the base for the target system comprising speaker identification, speech recognition and speaker adaptation.

4.1 Motivation

Initializing and adapting the codebooks of the automatic speech recognizer introduced in Sect. 2.5.3 is highly demanding since about 25,000 parameters are involved in adapting mean vectors.

The key issue is to limit the effective number of adaptation parameters in the case of limited data and to allow a more individual adjustment later on. Thus, a balanced adaptation strategy relying on short-term adaptation at the beginning and individual adaptation in the long-run as well as a smooth transition has to be implemented.

First, it is assumed that speaker identification and speech recognition deliver a reliable guess of the speaker identity and the transcription of the processed speech utterance. Error propagation due to maladaptations is neglected in this chapter and will be addressed later. The benefit of robust speaker adaptation for speech recognition will be demonstrated by experiments under the condition that the speaker identity is given.

T. Herbig, F. Gerl, and W. Minker: Self-Learning Speaker Identification, SCT, pp. 71–82.
springerlink.com

4.2 Algorithm

In Sect. 2.6 an overview of several well-known adaptation techniques which capture the voice characteristics of a particular speaker was given:

Short-term adaptation such as EV adaptation imposes a tied adaptation scheme for all Gaussian densities of a codebook. Long-term adaptation implemented by the MAP adaptation permits individual modifications in the case of a large data set. However, MAP adaptation is expected to be inefficient during the start or enrollment phase of a new speaker model as explained in Sect. 2.6.3. In Sect. 2.6.6 an optimal combination of short-term and long-term adaptation was discussed. However, the EMAP algorithm requires a costly matrix inversion and is considered to be too expensive for the purpose of this book.

In the following a simple interpolation of short-term adaptation realized by eigenvoices and long-term adaptation based on the MAP technique is investigated. Kuhn et al. [2000] present results mostly for supervised training for MAP using an EV estimate as prior. They do not report any tuning of the interpolation factor for the MAP estimate. This may explain their results for MLED=>MAP which performed worse on isolated letters compared to pure EVs in the unsupervised case. Since the transition from EV adaptation to MAP is crucial for the system presented in this work, it will be derived in closer detail. Furthermore, the problem of efficient data acquisition for continuous adaptation in an unsupervised system will be addressed. Details will be provided on how to handle new speaker specific data and how to delay speaker adaptation until a robust guess of the speaker identity is available. Robustness will be gained by the initial modeling of a speech recognizer based on SCHMMs.

Eigenvoices have shown a very robust behavior in the experiments which will be discussed later. Only a few parameters are required. Prior knowledge about speaker adaptation effects, gender, channel impulse responses and background noises contributes to the robustness of EV adaptation. The key issue of the suggested adaptation scheme is to start from the prior distribution of $\boldsymbol{\mu}_k^{\mathrm{EV}}$ and to derive the resulting data-driven adaptation strategy.

A two-stage procedure is applied which is comparable to the segmental MAP algorithm in Sect. 2.6.2. Speech decoding provides an optimal state alignment and determines the state dependent weights w_k^s of a particular codebook. Then the codebook is adapted based on the training data $\mathbf{x}_{1:T}$ of the target speaker i. In the following derivation the auxiliary function of the EM algorithm is extended by a term comprising prior knowledge:

$$Q_{\mathrm{MAP}}(\Theta_i, \Theta_0) = Q_{\mathrm{ML}}(\Theta_i, \Theta_0) + \log\left(p(\Theta_i)\right) \tag{4.1}$$

$$= \sum_{t=1}^{T} \sum_{k=1}^{N} p(k|\mathbf{x}_t, \Theta_0) \cdot \log\left(p(\mathbf{x}_t, k|\Theta_i)\right) + \log\left(p(\Theta_i)\right) \tag{4.2}$$

as given in (2.40) and (2.78). Only one iteration of the EM algorithm is calculated. The state variable s is omitted. The initial parameter set

$$\Theta_0 = \left\{w_1^0, \ldots, w_N^0, \boldsymbol{\mu}_1^0, \ldots, \boldsymbol{\mu}_N^0, \boldsymbol{\Sigma}_1^0, \ldots, \boldsymbol{\Sigma}_N^0\right\} \tag{4.3}$$

is given by the standard codebook. Since only mean vectors are adapted, the following notation is used for the speaker specific codebooks:

$$\Theta_i = \left\{w_1^0, \ldots, w_N^0, \boldsymbol{\mu}_1^i, \ldots, \boldsymbol{\mu}_N^i, \Sigma_1^0, \ldots, \Sigma_N^0\right\}. \tag{4.4}$$

Subsequently, the speaker index i is omitted. For reasons of simplicity, $\boldsymbol{\mu}_k$ and Σ_k denote the speaker specific mean vectors to be optimized and the covariance matrices of the standard codebook.

For the following equations it is assumed that each Gaussian distribution can be treated independently from the remaining distributions. Thus, the prior distribution of the mean vector $\boldsymbol{\mu}_k$ can be factorized or equivalently the logarithm is given by a sum of logarithms. A prior Gaussian distribution is assumed

$$\log\left(p(\boldsymbol{\mu}_1, \ldots, \boldsymbol{\mu}_N)\right) = \sum_{k=1}^{N} \log\left(\mathcal{N}\left\{\boldsymbol{\mu}_k | \boldsymbol{\mu}_k^{\mathrm{EV}}, \Sigma_k^{\mathrm{EV}}\right\}\right) \tag{4.5}$$

$$\begin{aligned} &= -\frac{1}{2}\sum_{k=1}^{N} \left(\boldsymbol{\mu}_k - \boldsymbol{\mu}_k^{\mathrm{EV}}\right)^T \cdot (\Sigma_k^{\mathrm{EV}})^{-1} \cdot \left(\boldsymbol{\mu}_k - \boldsymbol{\mu}_k^{\mathrm{EV}}\right) \\ &\quad - \frac{1}{2}\sum_{k=1}^{N} \log\left(|\Sigma_k^{\mathrm{EV}}|\right) - \frac{d}{2}\sum_{k=1}^{N} \log\left(2\pi\right) \end{aligned} \tag{4.6}$$

where the covariance matrix Σ_k^{EV} represents the uncertainty of the EV adaptation.

Q_{MAP} is maximized by taking the derivative with respect to the mean vector $\boldsymbol{\mu}_k$ and by calculating the corresponding roots $\boldsymbol{\mu}_k^{\mathrm{opt}}$. The derivation can be found in Sect. A.2. In this context, $\boldsymbol{\mu}_k^{\mathrm{EV}}$ is employed as prior information about the optimized mean vector $\boldsymbol{\mu}_k^{\mathrm{opt}}$. The optimization problem is solved by (A.26):

$$n_k \cdot \Sigma_k^{-1} \cdot \left(\boldsymbol{\mu}_k^{\mathrm{ML}} - \boldsymbol{\mu}_k^{\mathrm{opt}}\right) = (\Sigma_k^{\mathrm{EV}})^{-1} \cdot \left(\boldsymbol{\mu}_k^{\mathrm{opt}} - \boldsymbol{\mu}_k^{\mathrm{EV}}\right). \tag{4.7}$$

When n_k approaches infinity, only ML estimates are applied to adjust codebooks. In the learning phase which is characterized by limited training data the prior knowledge incorporated into the eigenvoices still allows efficient codebook initialization and continuous adaptation.

In the next step both sides of (4.7) are multiplied by Σ_k. Equation (4.7) can be rewritten by

$$\left(\Sigma_k \cdot (\Sigma_k^{\mathrm{EV}})^{-1} + n_k \cdot \mathrm{I}\right) \cdot \boldsymbol{\mu}_k^{\mathrm{opt}} = \Sigma_k \cdot (\Sigma_k^{\mathrm{EV}})^{-1} \cdot \boldsymbol{\mu}_k^{\mathrm{EV}} + n_k \cdot \boldsymbol{\mu}_k^{\mathrm{ML}} \tag{4.8}$$

where I denotes the identity matrix. The matrix $\Sigma_k \cdot (\Sigma_k^{\mathrm{EV}})^{-1}$ makes use of prior knowledge about statistical dependencies between the elements of each mean vector. However, the EV approach already contains prior knowledge as described in Sect. 2.6.5. It may be assumed that any matrix multiplication can be discarded. If the approximation

$$\Sigma_k \cdot (\Sigma_k^{\mathrm{EV}})^{-1} = \lambda \cdot \mathrm{I}, \qquad \lambda = \mathrm{const} \tag{4.9}$$

is used, the solution of the optimization problem becomes quite easy to interpret [Herbig et al., 2010c]:

$$\boldsymbol{\mu}_k^{\mathrm{opt}} = (1-\alpha_k) \cdot \boldsymbol{\mu}_k^{\mathrm{EV}} + \alpha_k \cdot \boldsymbol{\mu}_k^{\mathrm{ML}} \tag{4.10}$$

$$\alpha_k = \frac{n_k}{n_k + \lambda}. \tag{4.11}$$

Since this adaptation scheme interpolates EV and ML estimates within the MAP framework, it is called EV-MAP adaptation in the following.

When sufficient speaker specific data is available, the convex combination approaches ML estimates such as the MAP adaptation[1] described in Sect. 2.6.3.

Few data causes this adaptation scheme to resemble the EV estimate $\boldsymbol{\mu}_k^{\mathrm{EV}}$. The latter is decomposed into a linear combination of the eigenvoices as defined in (2.101). The corresponding weights are calculated according to (2.116). Only the sufficient statistics of the standard codebook

$$n_k = \sum_{t=1}^{T} p(k|\mathbf{x}_t, \Theta_0) \tag{4.12}$$

$$\boldsymbol{\mu}_k^{\mathrm{ML}} = \frac{1}{n_k} \sum_{t=1}^{T} p(k|\mathbf{x}_t, \Theta_0) \cdot \mathbf{x}_t \tag{4.13}$$

are required. However, nearly untrained Gaussian densities do not provide reliable ML estimates. In fact, this computation may lead to diverging values of $\boldsymbol{\mu}_k^{\mathrm{ML}}$ and may force extreme weighting factors in (2.116). This may reduce the performance of speaker identification and speech recognition. Thus, the ML estimates in (2.116) are replaced here by MAP estimates as given in (2.85). For sufficiently large n_k there is no difference since $\boldsymbol{\mu}_k^{\mathrm{MAP}}$ and $\boldsymbol{\mu}_k^{\mathrm{ML}}$ converge. If n_k is small for a particular Gaussian distribution, the MAP estimate guarantees that approximately the original mean vector is used $\boldsymbol{\mu}_k^{\mathrm{MAP}} \approx \boldsymbol{\mu}_k^{0}$. Those mean vectors have only a limited influence on the result of the set of linear equations in (2.116). Furthermore, the EV adaptation was only calculated in the experiments carried out when sufficient speech data ≥ 0.5 sec was available.

The ML estimates have to be continuously updated so that $\boldsymbol{\mu}_k^{\mathrm{EV}}$ and $\boldsymbol{\mu}_k^{\mathrm{opt}}$ can be computed. Since the ML estimates $\boldsymbol{\mu}_k^{\mathrm{ML}}$ are weighted sums of all observed feature vectors assigned to a particular Gaussian distribution, a

---

[1] In the following, the term MAP adaptation refers to the implementation discussed in Sect. 2.6.3. EV-MAP adaptation denotes the adaptation scheme explained in this chapter.

recursive computation is possible. The updated ML estimates $\boldsymbol{\mu}_k^{\mathrm{ML}}$ are weighted sums of the preceding estimates $\bar{\boldsymbol{\mu}}_k^{\mathrm{ML}}$ and an innovation term

$$n_k = \bar{n}_k + \sum_{t=1}^{T} p(k|\mathbf{x}_t, \Theta_0) \tag{4.14}$$

$$\boldsymbol{\mu}_k^{\mathrm{ML}} = \frac{\bar{n}_k}{n_k} \cdot \bar{\boldsymbol{\mu}}_k^{\mathrm{ML}} + \frac{1}{n_k} \cdot \sum_{t=1}^{T} p(k|\mathbf{x}_t, \Theta_0) \cdot \mathbf{x}_t \tag{4.15}$$

where $\bar{n}_k$ denotes the number of the assigned feature vectors up to the last utterance. This allows efficient long-term speaker tracking which will be discussed later. Speaker specific adaptation data can be collected before a guess of the speaker identity is available.

A concrete realization should apply an exponential weighting window for $\boldsymbol{\mu}_k^{\mathrm{ML}}$ and n_k. A permanent storage may be undesirable, since speaker characteristics significantly change over time[2] [Furui, 2009]. In this book no window is applied because time-variance is not a critical issue for the investigated speech database.

Finally, the solution of (4.8) may be discussed with respect to the EMAP solution in (2.122). Equation (4.8) is equivalent to

$$\begin{aligned} &\left(\Sigma_k \cdot (\Sigma_k^{\mathrm{EV}})^{-1} + n_k \cdot \mathrm{I}\right) \cdot \boldsymbol{\mu}_k^{\mathrm{opt}} \\ &= \left(\Sigma_k \cdot (\Sigma_k^{\mathrm{EV}})^{-1} + n_k \cdot \mathrm{I}\right) \cdot \boldsymbol{\mu}_k^{\mathrm{EV}} + n_k \cdot \left(\boldsymbol{\mu}_k^{\mathrm{ML}} - \boldsymbol{\mu}_k^{\mathrm{EV}}\right). \end{aligned} \tag{4.16}$$

Both sides are multiplied by $\left(\Sigma_k \cdot (\Sigma_k^{\mathrm{EV}})^{-1} + n_k \cdot \mathrm{I}\right)^{-1}$ to obtain the following term

$$\boldsymbol{\mu}_k^{\mathrm{opt}} = \boldsymbol{\mu}_k^{\mathrm{EV}} + n_k \cdot \left(\Sigma_k \cdot (\Sigma_k^{\mathrm{EV}})^{-1} + n_k \cdot \mathrm{I}\right)^{-1} \cdot \left(\boldsymbol{\mu}_k^{\mathrm{ML}} - \boldsymbol{\mu}_k^{\mathrm{EV}}\right) \tag{4.17}$$

which is equivalent to

$$\boldsymbol{\mu}_k^{\mathrm{opt}} = \boldsymbol{\mu}_k^{\mathrm{EV}} + n_k \cdot \Sigma_k^{\mathrm{EV}} \cdot \left(\Sigma_k + n_k \cdot \Sigma_k^{\mathrm{EV}}\right)^{-1} \cdot \left(\boldsymbol{\mu}_k^{\mathrm{ML}} - \boldsymbol{\mu}_k^{\mathrm{EV}}\right). \tag{4.18}$$

A comparison with (2.122) identifies a similar structure of both adaptation terms. The matrix Σ_k^{EV} corresponds to $\check{\Sigma}^{\mathrm{EMAP}}$. Despite these structural similarities, there is a basic difference between EMAP and the adaptation scheme proposed here. EMAP requires the computation and especially the matrix inversion in the supervector space. The prior knowledge is represented by $\check{\Sigma}^{\mathrm{EMAP}}$. Combining EV and ML estimates takes benefit from statistical bindings in the EV estimation process. However, this approach works in the feature space which is characterized by a lower dimensionality.

Botterweck [2001] describes a similar combination of MAP and EV estimates. The original equation is re-written by

[2] A study of intra-session and inter-session variability can be found by Godin and Hansen [2010].

$$\check{\boldsymbol{\mu}} = \check{\boldsymbol{\mu}}^0 + \check{\mathsf{P}} \cdot \left(\check{\boldsymbol{\mu}}^{\mathrm{MAP}} - \check{\boldsymbol{\mu}}^0\right) + \sum_{l=1}^{L} \alpha_l \cdot \check{\mathbf{e}}_l^{\mathrm{EV}} \tag{4.19}$$

to agree with the notation of Sect. 2.6. Using the MAP framework, the coefficients of the EVs α_l have to be determined by optimizing the auxiliary function of the EM algorithm. $\check{\boldsymbol{\mu}}^0$ represents all mean vectors of the standard codebook in supervector notation. The matrix $\check{\mathsf{P}}$ guarantees that the standard MAP estimate only takes effect in the directions orthogonal to all eigenvoices. Adaptation is performed in the supervector space. Due to the matrix multiplication it is more expensive than the solution derived above.

4.3 Evaluation

Several experiments have been conducted with different speaker adaptation methods to investigate the benefit which can be achieved by a robust speaker identification. The results are presented and discussed in detail.

First, the database is introduced and the composition of the selected subset is explained. This subset is subsequently used for all evaluations. The evaluation setup is described in more detail. It is employed to simulate realistic tasks for self-learning speaker identification and speech recognition. Finally, the results for closed-set experiments are presented.

4.3.1 Database

SPEECON[3] is an extensive speech database which was collected by a consortium comprising several industrial partners. The database comprises about 20 languages and is intended to support the development of voice-driven interfaces of consumer applications. Each language is represented by 600 speakers comprising 550 adult speakers. The recordings were done under realistic conditions and environments such as home, office, public places and in-car. Both read and spontaneous speech are covered.

In this book, a subset of the US-SPEECON database is used for evaluation. This subset comprises 73 speakers (50 male and 23 female speakers) recorded in an automotive environment. The Lombard effect is considered. The sound files were down-sampled from 16 kHz to 11.025 kHz. Only AKG microphone recordings were used for the experiments. Colloquial utterances with more than 4 words were removed to obtain a realistic command and control application. Utterances containing mispronunciations were also discarded. Digit and spelling loops were kept irrespective of their length. This results in at least 250 utterances per speaker which are ordered in the sequence of the recording session.

[3] The general description of the SPEECON database follows Iskra et al. [2002].

4.3.2 Evaluation Setup

In the following the configuration of the test set, baseline and speaker adaptation is explained. Furthermore, the evaluation measures for speech recognition and speaker identification are introduced.

Test Set

The evaluation was performed on 60 sets. Each set comprises 5 enrolled speakers[4] to ensure independence of the speaker set composition which is chosen randomly. The composition of female and male speakers in a group is not balanced. At the beginning of each speaker set an unsupervised enrollment takes place. It consists of 10 utterances for each speaker as shown in Fig. 4.1. In Chapter 5 and 6 closed-set and open-set scenarios will be discussed.

For closed sets the system has to identify a speaker from a list of enrolled speakers. Thus, the first two utterances of each new speaker are indicated during enrollment. For the remaining 8 utterances the system has to identify the speaker in an unsupervised manner. The corresponding speaker specific codebook has to compete against existing speaker models. In an open-set scenario the system has to decide when a new speaker model has to be added to the existing speaker profiles.

After this enrollment at least 5 utterances are spoken before the speaker changes. From one utterance to the next the probability of a speaker change is approximately 10 %. The order of the speakers in each set are chosen randomly. For each set all utterances of each speaker are used only once. Thus, every set provides $1,250$ utterances originating from 5 speakers. This results in $75,000$ utterances used for evaluation. The speaker specific codebooks are continuously adapted. The configuration of the test investigates the robustness of the start phase, error propagation and long-term speaker adaptation. No additional information concerning speaker changes and speaker identities will be given in Chapter 5 and 6.

Baseline

The speech recognizer without any speaker adaptation or speaker profiles acts as the baseline for all subsequent experiments. The speech recognizer applies grammars for digit and spelling loops as well as for numbers. Finally, a grammar was generated which contains all remaining utterances ($\approx 2,000$ command and control utterances). For testing the corresponding grammar was selected prior to each utterance. The speech recognizer was trained as described by Class et al. [1993].

---

[4] When speaker sets with 10 enrolled speakers are examined in Chapter 6, another evaluation set comprising 30 different speaker sets is used.

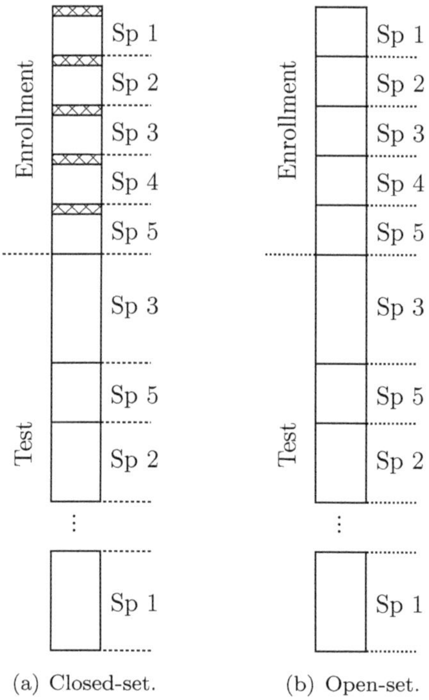

Fig. 4.1 Block diagram of the setup which is used for all experiments of this book. The setup consists of two blocks. First, a start phase is used in which new speaker specific codebooks are initialized. However, the speaker identity is only given in the supervised parts (diagonal crossed boxes) of Fig. 4.1(a). For the remaining utterances speaker identification is delegated to the speech recognizer or a common speaker identification system. Then speakers are randomly selected. At least 5 utterances are spoken by each speaker before a speaker turn takes place.

Speaker Adaptation

In an off-line training the main directions of speaker variability in the feature space were extracted. For this purpose codebooks were trained by MAP adaptation for a large set of speakers of the USKCP[5] development database. The mean vectors of each codebook were stacked into long vectors. The eigenvoices were calculated using PCA.

To simulate continuous speaker adaptation without speaker identification, short-term adaptation is implemented by a modified EV approach. A smoothing of the weighting factors $\boldsymbol{\alpha}$ is applied by introducing an exponential weighting window in (4.14) and (4.15). It guarantees that speaker changes are

[5] The USKCP is a speech database internally collected by TEMIC Speech Dialog Systems, Ulm, Germany. The USKCP comprises command and control utterances recorded in an automotive environment. The language is US-English.

captured within approximately 5 utterances if no speaker identification is employed. However, speech recognition is still affected by frequent speaker changes.

Evaluation Measures

Speech recognition is evaluated by the *Word Accuracy* (WA)

$$\mathrm{WA} = 100 \cdot \left(1 - \frac{W_{\mathrm{S}} + W_{\mathrm{I}} + W_{\mathrm{D}}}{W}\right) \%. \tag{4.20}$$

W represents the number of words in the reference string and W_{S}, W_{I} and W_{D} denote the number of substituted, inserted and deleted words in the recognized string [Boros et al., 1996]. Speaker identification is evaluated by the identification rate. The typical error is given by a confidence interval assuming independent data sets. Only the intervals of the best and the worst result are given. When a test is repeated on an identical test set, a paired difference test may be more appropriate to evaluate the statistical significance. In Sect. A.3 the evaluation measures are described in detail.

4.3.3 Results

Before a realistic closed-set scenario is examined in detail, an upper bound for unsupervised speaker adaptation is investigated [Herbig et al., 2010c]. It determines the optimal WA of the speech recognizer which can be achieved by a perfect speaker identification combined with several adaptation strategies. In addition, feature extraction is externally controlled so that energy normalization and mean subtraction are updated or reset before the next utterance is processed. Nevertheless, unsupervised speaker adaptation has to handle errors of speech recognition. Short-term adaptation and the speaker adaptation scheme introduced in Sect. 4.2 are compared to the baseline system.

The tuning parameter λ in (4.11) is evaluated for specific values. For $\lambda \approx 0$ the adaptation relies only on ML estimates. In contrast, $\lambda \to \infty$ is the reverse extreme case since EV estimates are used irrespective of the amount of speaker specific data. The latter case is expected to limit the improvement of word accuracy. In addition, the performance of MAP adaptation is tested for dedicated values of the tuning parameter η used in (2.84) and (2.85).

In Table 4.1 and Fig. 4.2 the speech recognition rates are compared for the complete evaluation test set under the condition that the speaker identity is known. The results show a significant improvement of WA when the EV-MAP speaker adaptation is compared to baseline and short-term adaptation. Only about 6 % relative error rate reduction is achieved by short-term adaptation compared to the baseline. EV adaptation combined with perfect speaker identification yields about 20 % relative error rate reduction. 88.94 % WA

Table 4.1 Comparison of different adaptation techniques under the condition of **perfect speaker identification**. The baseline is the speaker independent speech recognizer without any speaker adaptation. Short-term adaptation without speaker identification is realized by EV adaptation with an exponential weighting window. This adaptation is based only on a few utterances. EV-MAP speaker adaptation is tested for several dedicated values of the parameter λ. Speaker pools with 5 speakers are investigated. Table is taken from [Herbig et al., 2010c].

Speaker adaptation	WA [%]
Baseline	85.23
Short-term adaptation	86.13
EV-MAP adaptation	
ML ($\lambda \approx 0$)	88.44
$\lambda = 4$	88.87
$\lambda = 8$	88.86
$\lambda = 12$	**88.94**
$\lambda = 16$	88.86
$\lambda = 20$	88.90
EV ($\lambda \to \infty$)	88.19
MAP adaptation	
$\eta = 4$	88.62
$\eta = 8$	88.62
$\eta = 12$	88.57
Typical errors	
min	± 0.22
max	± 0.25

and 25 % relative error rate reduction with respect to the speaker independent baseline are achieved using $\lambda = 12$. Obviously, λ can be selected in a relatively wide range without negatively affecting the performance of the speech recognizer. The results for pure EV or MAP adaptation are worse as expected. MAP adaptation yields better results than the ML or EV approach [Herbig et al., 2010c].

In Table 4.2 the evaluation is split into several evaluation bands representing different adaptation stages. The effect of speaker adaptation is evaluated for weakly, moderately and extensively trained codebooks. Since the evaluation is done on different speech utterances the recognition rates and improvements are only comparable within each evaluation band.

The implementations which interpolate EV and ML estimates yield consistent results for all evaluation bands. On the evaluation bands I and III the main difference between ML estimates and EV adaptation can be seen. On the first utterances EV adaptation is superior to ML estimates which will become obvious in the following chapters. In the long-run MAP or ML estimates allow more individual codebook adaptation. They outperform the EV approach in the long run, but suffer from poor performance during the

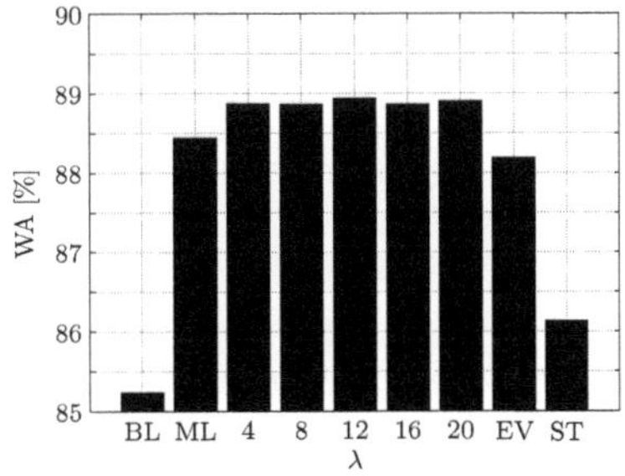

Fig. 4.2 Results for speaker adaptation with predefined speaker identity. Furthermore, the speaker independent baseline (BL) and short-term adaptation (ST) are depicted.

Table 4.2 Comparison of different adaptation techniques under the condition of **perfect speaker identification**. The evaluation is split into several evaluation bands according to the adaptation stage in terms of recognized utterances. Speech recognition results are presented for several evaluation bands - I = [1; 50[, II = [50; 100[, III = [100; 250] utterances.

Speaker adaptation	I	II	III
	WA [%]	WA [%]	WA [%]
Baseline	84.51	86.84	85.02
Short-term adaptation	85.89	87.22	85.87
EV-MAP adaptation			
ML ($\lambda \approx 0$)	87.77	89.17	88.50
$\lambda = 4$	88.43	89.67	**88.80**
$\lambda = 8$	**88.57**	**89.97**	88.62
$\lambda = 12$	88.51	89.95	88.79
$\lambda = 16$	88.35	89.91	88.74
$\lambda = 20$	88.37	89.83	**88.84**
EV ($\lambda \to \infty$)	87.99	89.03	**88.00**
MAP adaptation			
$\eta = 4$	**87.98**	89.36	88.67
$\eta = 8$	87.69	**89.54**	88.73
$\eta = 12$	87.56	89.42	**88.75**
Typical errors			
min	±0.51	±0.48	±0.29
max	±0.58	±0.54	±0.33

learning phase. Therefore the EV approach performs worse than ML estimates on the complete test set.

It should be emphasized that this experiment was conducted under optimal conditions. In the following chapters it will be referred to this experiment as an upper bound for tests under realistic conditions where the speaker has to be identified in an unsupervised way. The ML approach will suffer from poor performance when the number of utterances is smaller than ≈ 40 as will be shown in Chapter 6. In this case the recognition accuracy is expected to be limited in the long run.

4.4 Summary

A flexible adaptation scheme was presented and discussed. Speaker specific codebooks can be initialized and continuously adapted based on the standard codebook. The algorithm provides a data-driven smooth transition between globally estimated mean vectors and locally optimized mean vectors permitting individual adjustment. Computational complexity is negligible compared to the EMAP algorithm. ML estimates do not have to be re-calculated as being already available from the EV adaptation. The interpolation of EV and ML estimates only requires to calculate the weighting factors α_k and the sum of the weighted estimates for each Gaussian density according to (4.10).

When the system has perfect knowledge of the identity of the current speaker, a relative error rate reduction of 25 % can be achieved via the adaptation techniques described here. Both the MAP and EV adaptation described in Sect. 2.6.3 and Sect. 2.6.5 have performed worse than the combined approach since a fast convergence during the initialization and an individual training on extensive adaptation data are required as well as an optimal smooth transition.

The conditions of this experiment may be realistic for some systems: A system may ask each speaker to identify himself when he starts a speaker turn or when the system is reset. However, this assumption is expected to limit the application of some adaptation techniques to use cases where the speaker has to be identified in an unsupervised manner. Thus, several techniques for speaker identification are discussed in the next chapters to approach this upper bound.

5

Unsupervised Speech Controlled System with Long-Term Adaptation

Chapters 2 and 3 provided a general overview of problems related to the scope of this book. The fundamentals and existing solutions as known from literature were discussed in detail.

In the preceding chapter a combination of short-term and long-term adaptation was examined. Speaker profiles can be initialized and continuously adapted under the constraint that the speaker has to be known a priori. The experiments carried out showed a significant improvement for speech recognition if speaker identity is known or can be at least reliably estimated.

In this chapter a flexible and efficient technique for speaker specific speech recognition is presented. This technique does not force the user to identify himself. Speaker variabilities are handled by an unsupervised speech controlled system comprising speech recognition and speaker identification.

The discussion starts with the problem to be solved and motivates the new approach. Then a unified speaker identification and speech recognition method is presented to determine the speaker's identity and to decode the speech utterance simultaneously. A unified modeling is described which handles both aspects of speaker and speech variability. The basic architecture of the target system including the modules speaker identification, speech recognition and speaker adaptation is briefly discussed.

A reference implementation is then described. It comprises a standard speaker identification technique in parallel to the speaker specific speech recognition as well as speaker adaptation of both statistical models. It is used as a reference implementation for the subsequent experiments.

Both implementations are evaluated for an in-car application. Finally, both systems are discussed with respect to the problem of this book and an extension of the speaker identification method is motivated.

5.1 Motivation

In the preceding chapters the effects of speaker variability on speech recognition were described and several techniques which solve the problem of speaker

T. Herbig, F. Gerl, and W. Minker: Self-Learning Speaker Identification, SCT, pp. 83–113.
springerlink.com

tracking and unsupervised adaptation were outlined. Each of the approaches described covers a part of the use case in this book.

The main goal is now to construct a complete system which includes speaker identification, speech recognition and speaker adaptation. Ideally, it shall be operated in a completely unsupervised manner so that the user can interact with the system without any additional intervention.

Simultaneous speaker identification and speech recognition can be described as a MAP estimation problem

$$\{\mathcal{W}_{1:N_\mathrm{W}}^{\mathrm{MAP}}, i^{\mathrm{MAP}}\} = \arg \max_{\mathcal{W}_{1:N_\mathrm{W}}, i} \{p(\mathcal{W}_{1:N_\mathrm{W}}, i | \mathbf{x}_{1:T})\} \tag{5.1}$$

$$= \arg \max_{\mathcal{W}_{1:N_\mathrm{W}}, i} \{p(\mathcal{W}_{1:N_\mathrm{W}} | i, \mathbf{x}_{1:T}) \cdot p(i|\mathbf{x}_{1:T})\} \tag{5.2}$$

as found by Nakagawa et al. [2006]. The parameter set Θ_i of the speaker specific codebook is omitted. The MAP criterion is applied to the joint posterior probability of the spoken phrase $\mathcal{W}_{1:N_\mathrm{W}}$ and speaker identity i given the observed feature vectors $\mathbf{x}_{1:T}$.

An approach is subsequently presented which is characterized by a mutual benefit between speaker identification and speech recognition. Computational load and latency are reduced by an on-line estimation of the speaker identity. For each feature vector the most probable speaker is determined. The associated speaker profile enables speech decoding. An efficient implementation with respect to computational complexity and memory consumption is targeted.

Models used in speaker change detection, speaker identification and speech recognition are of increasing complexity. BIC is based on single Gaussian distributions whereas the speech recognizer introduced in Sect. 2.5.3 is realized by a mixture of Gaussian distributions with an underlying Markov model.

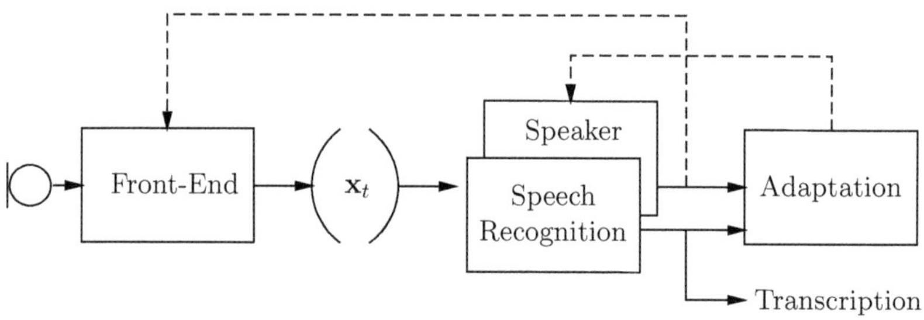

Fig. 5.1 Unified speaker identification and speaker specific speech recognition. One feature extraction and only one statistical model are employed for speaker identification and speech recognition. The estimated speaker identity and the transcription of the spoken phrase are used to initialize and continuously adapt speaker specific models. Speaker identification and speech recognition of successive utterances are enhanced.

The goal is to show in this book that the problem of speaker variability can be solved by a compact statistical modeling which allows unsupervised simultaneous speaker identification and speech recognition in a self-evolving system [Herbig and Gerl, 2009; Herbig et al., 2010d].

The subsequent sections describe a unified statistical model for both tasks as an extension of the speaker independent speech recognizer which was introduced in Sect. 2.5.3. It is shown how a robust self-learning speaker identification with enhanced speech recognition can be achieved for a closed-set scenario. The system structure is depicted in Fig. 5.1.

5.2 Joint Speaker Identification and Speech Recognition

In the preceding chapter an appropriate speaker adaptation technique was introduced to initialize and adapt codebooks of a speech recognizer. However, the derivation of the adaptation scheme was done under the condition that speaker identity and the spoken text are known. Now speaker specific speech recognition and speaker identification have to be realized.

The implementation of an automated speech recognizer presented in Sect. 2.5.3 is based on a speaker independent codebook. As shown in Fig. 2.11 the feature vectors extracted by the front-end are compared with a speaker independent codebook. For each time instance the likelihood $p(\mathbf{x}_t|k, \Theta_0)$ is computed for each Gaussian distribution with index k. The resulting vector $\mathbf{q}_t$ given by (2.67) and (2.68) comprises all likelihoods and is used for the subsequent speech decoding.

When the speaker's identity is given, speaker adaptation can provide individually adapted codebooks which increase the speech recognition rate for each speaker. Therefore, the speech recognizer has to be able to handle several speaker profiles in an efficient way.

In Fig. 5.2 a possible realization of an advanced speaker specific speech recognizer is shown. The speech recognizer is extended by N_{Sp} speaker specific codebooks which are operated in parallel to the speaker independent codebook. This approach allows accurate speech modeling for each speaker. However, the parallel computation significantly increases the computational load since the speech decoder has to process $N_{\mathrm{Sp}} + 1$ data streams in parallel. The Viterbi algorithm is more computationally complex than a codebook

Fig. 5.2 Block diagram for a speaker specific ASR system. N_{Sp} speaker specific codebooks and speech decoders are operated in parallel to generate a transcription for each speaker profile.

selection strategy on a frame-level, especially for a large set of speaker specific codebooks operated in parallel.

Thus, it seems to be advantageous to continuously forward only the result $\mathbf{q}_t^i$ of the codebook belonging to the current speaker i. A solution without parallel speech decoding for each codebook is targeted. Such an approach, however, requires the knowledge of the current speaker identity for an optimal codebook selection.

The computational complexity can be significantly decreased by a two-stage approach. The solution of (5.2) can be approximated by the following two steps. First, the most probable speaker is determined by

$$i^{\mathrm{MAP}} = \arg\max_{i} \left\{ p(i|\mathbf{x}_{1:T}) \right\} \tag{5.3}$$

which can be implemented by standard speaker identification methods. In the second step, the complete utterance can be reprocessed. The corresponding codebook can then be employed in the speech decoder to generate a transcription. The MAP estimation problem

$$\mathcal{W}_{1:N_{\mathrm{W}}}^{\mathrm{MAP}} = \arg\max_{\mathcal{W}_{1:N_{\mathrm{W}}}} \left\{ p(\mathcal{W}_{1:N_{\mathrm{W}}}|\mathbf{x}_{1:T}, i^{\mathrm{MAP}}) \right\} \tag{5.4}$$

is then solved given the speaker identity i^{MAP}. For example, the most likely word string can be determined by the Viterbi algorithm which was described in Sect. 2.5.2. The disadvantages of such an approach are the need to buffer the entire utterance and reprocessing causing latency.

Thus, speaker identification is implemented here on two time scales. A fast but probably less confident identification selects the optimal codebook on a frame-level and enables speaker specific speech recognition. In addition, speaker identification on an utterance-level determines the current speaker and provides an improved guess of the speaker identity for speaker adaptation. The latter is performed after speech decoding. The goal is to identify the current user and to decode the spoken phrase in only one iteration [Herbig et al., 2010d].

5.2.1 Speaker Specific Speech Recognition

An appropriate speech model has to be selected on a frame-level for speech decoding. Since high latencies of the speech recognition result and multiple recognitions have to be avoided, the speech recognizer has to be able to switch rapidly between speaker profiles at the risk of not always applying the correct speech model.

On the other hand, if the speech recognizer selects an improper speaker specific codebook because two speakers yield similar pronunciation patterns, the effect on speech recognition should be acceptable and no serious increase in the error rate is expected. False speaker identifications lead to mixed codebooks when the wrong identity is employed in speaker adaptation. It seems to

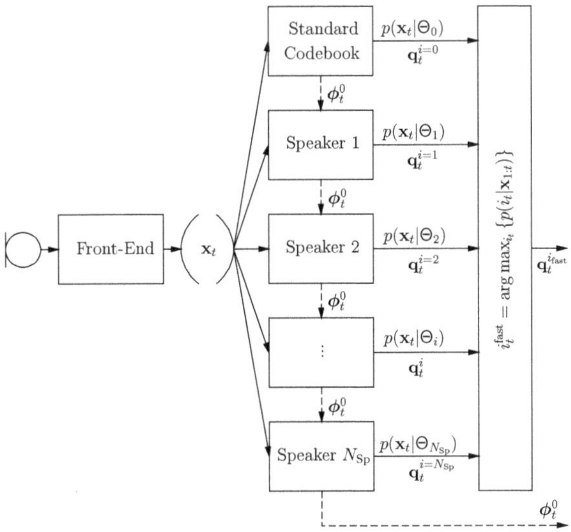

Fig. 5.3 Implementation of the codebook selection for speaker specific speech recognition. Appropriate codebooks are selected and the corresponding soft quantization $\mathbf{q}_t^{i_{\text{fast}}}$ is forwarded to speech decoding. Figure is taken from [Herbig et al., 2010d].

be evident that this kind of error can have a more severe impact on the long-term stability of the speech controlled system than an incorrect codebook selection for a single utterance.

Subsequently, a strategy for codebook selection is preferred. It is oriented at the match between codebooks and the observed feature vectors $\mathbf{x}_t$. Since a codebook represents the speaker's pronunciation characteristics, this decision is expected to be correlated with the speaker identification result.

Class et al. [2003] describe a technique for automatic speaker change detection which is based on the evaluation of several speaker specific codebooks. This approach is characterized by low computational complexity. It is extended in this book by speaker specific speech recognition and simultaneous speaker tracking.

Fig. 5.3 displays the setup of parallel codebooks comprising one speaker independent and several speaker specific codebooks. The standard codebook with index $i = 0$ is used to compute the match $\mathbf{q}_t^0$ between this codebook and the feature vector $\mathbf{x}_t$. To avoid latencies caused by speech decoding, only codebooks are investigated and the underlying Markov models are not used. Therefore, the statistical models are reduced in this context to GMMs. Even in the case of an SCHMM the weighting factors $w_k^{s_t}$ are state dependent. The state alignment is obtained from the acoustic models during speech decoding which requires the result of the soft quantization $\mathbf{q}_t^i$. The codebooks are further reduced to GMMs with uniform weighting factors

$$w_k = \frac{1}{N} \tag{5.5}$$

for reasons of simplicity [Herbig et al., 2010d].

Even though the parallel speech decoding of $N_{\mathrm{Sp}}+1$ codebooks is avoided, the computational load for the evaluation of $N_{\mathrm{Sp}}+1$ codebooks on a frame-level can be undesirable for embedded systems. Thus, a pruning technique is applied to the likelihood computation of speaker specific codebooks. Only those Gaussian densities are considered which generate the N_{b} highest likelihoods $p(\mathbf{x}_t|k, \Theta_0)$ for the standard codebook. For $N_{\mathrm{b}} = 10$ no eminent differences between the resulting likelihood

$$p(\mathbf{x}_t|\Theta_i) = \frac{1}{N} \sum_{k=1}^{N} \mathcal{N}\left\{\mathbf{x}_t|\boldsymbol{\mu}_k^i, \boldsymbol{\Sigma}_k^0\right\} \tag{5.6}$$

and

$$p(\mathbf{x}_t|\Theta_i) \approx \frac{1}{N} \sum_{k \in \boldsymbol{\phi}_t^0} \mathcal{N}\left\{\mathbf{x}_t|\boldsymbol{\mu}_k^i, \boldsymbol{\Sigma}_k^0\right\} \tag{5.7}$$

have been observed in the experiments which will be discussed later [Herbig et al., 2010d]. Subsequently, the vector

$$\boldsymbol{\phi}_t^0 = \left(k_{t,1}^0, \ldots, k_{t,N_{\mathrm{b}}}^0\right)^T \tag{5.8}$$

represents the indices of the associated N_{b} Gaussian densities relevant for the likelihood computation of the standard codebook. Speaker specific codebooks emerge from the standard codebook and speaker adaptation always employs the feature vector assignment of the standard codebook as described in Sect. 4.2. Thus, the subset $\boldsymbol{\phi}_t^0$ can be expected to be a reasonable selection for all codebooks. Each codebook generates a separate vector

$$\mathbf{q}_t^i \propto \left(p(\mathbf{x}_t|k = \phi_{t,1}^0, \Theta_i), \ldots, p(\mathbf{x}_t|k = \phi_{t,N_{\mathrm{b}}}^0, \Theta_i)\right)^T \tag{5.9}$$

by evaluating this subset $\boldsymbol{\phi}_t^0$. The complexity of likelihood computation is reduced because only $N + N_{\mathrm{Sp}} \cdot N_{\mathrm{b}}$ instead of $N + N_{\mathrm{Sp}} \cdot N$ Gaussian densities have to be considered. Finally, each codebook generates a vector $\mathbf{q}_t^i$ and a scalar likelihood $p(\mathbf{x}_t|\Theta_i)$ which determines the match between codebook and observation.

A decision logic can follow several strategies to select an appropriate speaker specific codebook. Only the corresponding soft quantization $\mathbf{q}_t^{i_{\mathrm{fast}}}$ is further processed by the speech decoder. Subsequently, three methods are described:

One possibility may be a decision on the actual likelihood value $p(\mathbf{x}_t|\Theta_i)$. The ML criterion can be applied so that in each time instance the codebook is selected with the highest likelihood value

$$i_t^{\text{fast}} = \arg\max_i \left\{ p(\mathbf{x}_t|\Theta_i) \right\}. \tag{5.10}$$

The optimization problem for speech recognition can then be approximated by

$$\mathcal{W}_{1:N_\text{W}}^{\text{MAP}} = \arg\max_{\mathcal{W}_{1:N_\text{W}}} \left\{ p(\mathcal{W}_{1:N_\text{W}}|\mathbf{x}_{1:T}, i_{1:T}^{\text{fast}}) \right\} \tag{5.11}$$

where $i_{1:T}^{\text{fast}}$ describes the codebook selection given by the corresponding $\mathbf{q}_{1:T}$.

This criterion has the drawback that the codebook selection may change frame by frame. In the experiments carried out an unrealistic high number of speaker turns could be observed even within short utterances. This implementation seems to be inappropriate since the speech decoder is applied to a series of $\mathbf{q}_t$ originating from different codebooks. The speech recognizer has to be able to react quickly to speaker changes, e.g. from male to female, especially when no re-processing is allowed. However, an unrealistic number of speaker turns on a single utterance has to be suppressed.

A more advanced strategy for a suitable decision criterion might be to apply an exponential weighting window. Again, the iid assumption is used for reasons of simplicity. The sum of the log-likelihood values enables standard approaches for speaker identification as described in Sect. 5.2.2. A local speaker identification may be implemented by

$$l_t^i = l_{t-1}^i \cdot \alpha + \log\left(p(\mathbf{x}_t|\Theta_i)\right) \cdot (1-\alpha), \quad 1 < t \leq T,\ 0 < \alpha < 1 \tag{5.12}$$

$$l_1^i = p(\mathbf{x}_1|\Theta_i) \tag{5.13}$$

where l_t^i denotes the local log-likelihood of speaker i. Codebook selection may then be realized by

$$i_t^{\text{fast}} = \arg\max_i \left\{ l_t^i \right\} \tag{5.14}$$

and speech recognition can be described again by (5.11).

Due to the knowledge about the temporal context, an unrealistic number of speaker turns characterized by many switches between several codebooks can be suppressed. However, the identification accuracy is expected to be limited by the exponential weighting compared to (2.46).

Alternatively, a decision logic based on Bayes' theorem may be applied in analogy to the forward algorithm introduced in Sect. 2.5.2. The posterior probability $p(i_t|\mathbf{x}_{1:t})$ is estimated for each speaker i given the entire history of observations $\mathbf{x}_{1:t}$ of the current utterance. The posterior probability $p(i_t|\mathbf{x}_{1:t})$ decomposes into the likelihood $p(\mathbf{x}_t|i_t)$ and the predicted posterior probability $p(i_t|\mathbf{x}_{1:t-1})$:

$$p(i_t|\mathbf{x}_{1:t}) = \frac{p(\mathbf{x}_t|i_t, \mathbf{x}_{1:t-1}) \cdot p(i_t|\mathbf{x}_{1:t-1})}{p(\mathbf{x}_t|\mathbf{x}_{1:t-1})}, \quad t > 1 \tag{5.15}$$

$$\propto p(\mathbf{x}_t|i_t) \cdot p(i_t|\mathbf{x}_{1:t-1}) \tag{5.16}$$

$$p(i_1|\mathbf{x}_1) \propto p(\mathbf{x}_1|i_1) \cdot p(i_1). \tag{5.17}$$

The likelihood $p(\mathbf{x}_t|i_t, \mathbf{x}_{1:t-1})$ represents new information about the speaker identity given an observed feature vector $\mathbf{x}_t$ and is regarded independently from the feature vector's time history $\mathbf{x}_{1:t-1}$. The initial distribution $p(i_1)$ allows for a boost of a particular speaker identity. For convenience, the parameter set Θ is omitted here. The predicted posterior distribution may be given by

$$p(i_t|\mathbf{x}_{1:t-1}) = \sum_{i_{t-1}=1}^{N_{\mathrm{Sp}}} p(i_t|i_{t-1}) \cdot p(i_{t-1}|\mathbf{x}_{1:t-1}), \; t > 1. \tag{5.18}$$

This approach yields the advantage of normalized posterior probabilities instead of likelihoods. The codebook with maximal posterior probability is selected according to the MAP criterion:

$$i_t^{\mathrm{fast}} = \arg\max_i \left\{ p(i_t|\mathbf{x}_{1:t}) \right\}. \tag{5.19}$$

The transcription can be determined in a second step by applying the MAP criterion

$$\boldsymbol{\mathcal{W}}_{1:N_{\mathrm{W}}}^{\mathrm{MAP}} = \arg\max_{\boldsymbol{\mathcal{W}}_{1:N_{\mathrm{W}}}} \left\{ p(\boldsymbol{\mathcal{W}}_{1:N_{\mathrm{W}}}|\mathbf{x}_{1:T}, i_{1:T}^{\mathrm{fast}}) \right\} \tag{5.20}$$

which can be considered as an approximation of (5.2).

The standard codebook is always evaluated in parallel to all speaker specific codebooks to ensure that the speech recognizer's performance does not decrease if none of the speaker profiles is appropriate, e.g. during the initialization of a new codebook.

Since Markov models are not considered here, a smoothing of the posterior may be employed to prevent an instantaneous drop or rise, e.g. caused by background noises during short speech pauses.

This Bayesian framework yields the advantage that prior knowledge about a particular speaker, e.g. from the last utterance, can be easily integrated to stabilize codebook selection at the beginning of an utterance. A smoothed likelihood suffers from not normalized scores which makes it more complicated to determine an initial likelihood value for a preferred or likely speaker. Here the prior probability $p(i_1)$ can be directly initialized. For example, the discrete decision i_0 from the last utterance and the prior speaker change probability p_{Ch} may be used for initialization, e.g. given by

$$p(i_1) \propto \delta_{\mathrm{K}}(i_1, i_0) \cdot (1 - p_{\mathrm{Ch}}) + (1 - \delta_{\mathrm{K}}(i_1, i_0)) \cdot p_{\mathrm{Ch}}. \tag{5.21}$$

In summary, this representation is similar to a CDHMM. The states of the Markov model used for codebook selection are given by the identities of the enrolled speakers. The codebooks represent the pronunciation characteristics of particular speakers. Subsequently, only this statistical model is used for codebook selection.

5.2.2 Speaker Identification

In the preceding section codebook selection was implemented by likelihood computation for each codebook and applying Bayes' theorem. However, the techniques discussed in Sect. 5.2.1 should be viewed as a local speaker identification. Since speaker adaptation is calculated after speech decoding, a more robust speaker identification can be used on an utterance-level. Error propagation due to maladaptation seems to be more severe than an improper codebook selection on parts of an utterance.

The experiments discussed later clearly show the capability of speaker specific codebooks to distinguish different speakers. Furthermore, codebooks are expected to provide an appropriate statistical model of speech and speaker characteristics since HMM training includes more detailed knowledge of speech compared to GMMs. Thus, the key idea is to use the codebooks to decode speech utterances and to identify speakers in a single step as shown in Fig. 5.4.

Each speaker specific codebook can be evaluated in the same way as common GMMs for speaker identification. Only the simplification of equal weighting factors w_k is imposed for the same reasons as in Sect. 5.2.1. The likelihood computation can be further simplified by the iid assumption. The following notation is used.

The log-likelihood $\mathcal{L}_u^i$ denotes the accumulated log-likelihood of speaker i and the current utterance with index u. Each utterance is represented by its feature vector sequence $\mathbf{x}_{1:T_u}$. The log-likelihood is normalized by the length T_u of the recorded utterance which results in an averaged log-likelihood per frame

$$\mathcal{L}_u^i = \frac{1}{T_u} \log \left(p(\mathbf{x}_{1:T_u} | \Theta_i) \right) = \frac{1}{T_u} \sum_{t=1}^{T_u} \log \left(\sum_{k \in \phi_t^0} \mathcal{N} \left\{ \mathbf{x}_t | \boldsymbol{\mu}_k^i, \boldsymbol{\Sigma}_k^0 \right\} \right) \tag{5.22}$$

in analogy to Markov and Nakagawa [1996]. The weighting factors $w_k = \frac{1}{N}$ are omitted, for convenience.

However, this implementation of speaker identification suffers from speech pauses and garbage words which do not contain speaker specific information. The speech recognition result may be employed to obtain a more precise speech segmentation. Since speaker identification is enforced after speech decoding, the scalar likelihood scores $p(\mathbf{x}_t | \Theta_i)$ can be temporarily buffered. The accumulated log-likelihood

$$\mathcal{L}_u^i = \frac{1}{\#\boldsymbol{\Phi}} \sum_{t \in \boldsymbol{\Phi}} \log \left(\sum_{k \in \phi_t^0} \mathcal{N} \left\{ \mathbf{x}_t | \boldsymbol{\mu}_k^i, \boldsymbol{\Sigma}_k^0 \right\} \right) \tag{5.23}$$

is calculated for each speaker as soon as a precise segmentation is available. The set $\boldsymbol{\Phi}$ contains all time instances which are considered as speech. $\#\boldsymbol{\Phi}$

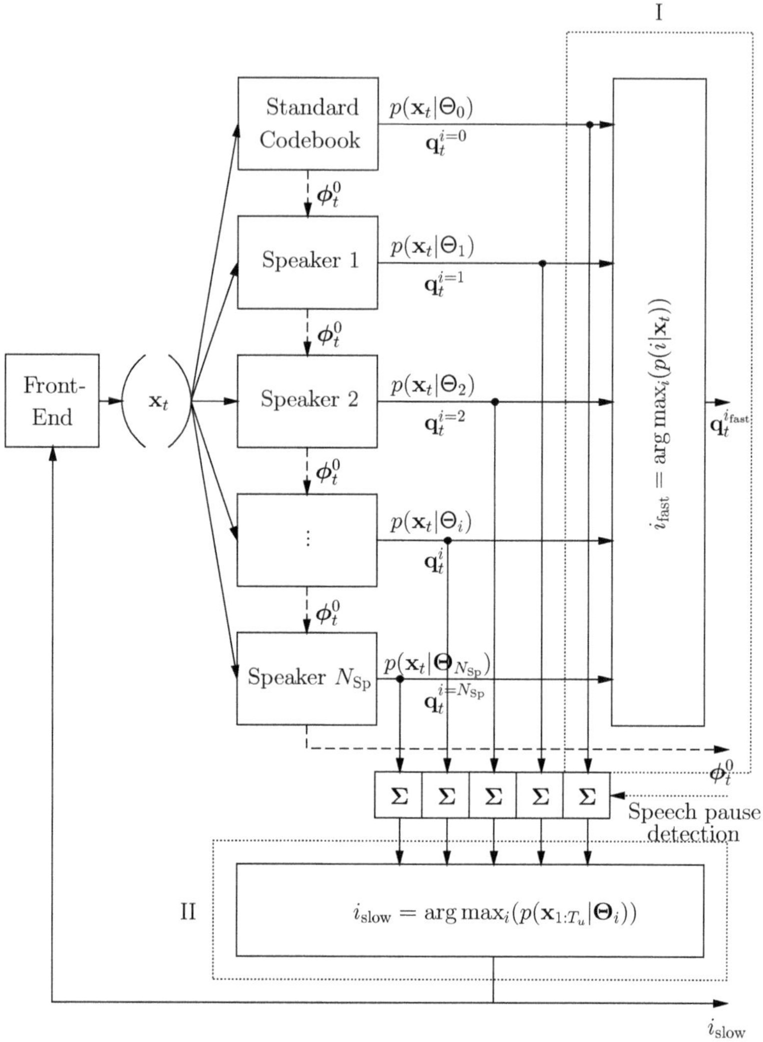

Fig. 5.4 Joint speaker identification and speech recognition. Each feature vector is evaluated by the standard codebook which determines the subset ϕ_t^0 comprising the N_b relevant Gaussian densities. Each speaker specific codebook is evaluated on the corresponding Gaussian densities given by ϕ_t^0. $p(\mathbf{x}_t|\Theta_0)$ and $\mathbf{q}_t^i$ are employed for speaker identification and speech recognition, respectively. Part I and II denote the codebook selection for speaker specific speech recognition as depicted in Fig. 5.3 and speaker identification on an utterance-level, respectively.

denotes the corresponding number of feature vectors containing purely speech. The modified log-likelihood ensures a relatively robust estimate

$$i_u^{\text{slow}} = \arg\max_i \left\{ \mathcal{L}_u^i \right\}. \tag{5.24}$$

Conventional speaker identification based on GMMs without additional speech recognition cannot prevent that neither training, adaptation nor test situations contain garbages or speech pauses.

To obtain a strictly unsupervised speech controlled system, the detection of new users is essential. It can be easily implemented by the following threshold decision

$$\mathcal{L}_u^i - \mathcal{L}_u^0 < \theta_{\text{th}}, \quad \forall i, \tag{5.25}$$

similar to Fortuna et al. [2005]. For example, the best in-set speaker is determined according to (5.24). The corresponding log-likelihood ratio is then tested for an out-of-set speaker.

5.2.3 System Architecture

Now all modules are available to construct a first complete system which is able to identify the speaker and to decode an utterance in one step. Speaker specific codebooks are initialized and continuously adapted as introduced in the preceding chapter. In Fig. 5.5 the block diagram of the system is depicted.

It is assumed that users have to press a push-to-talk button before starting to enter voice commands. Thus, speech segmentation is neglected since at least a rough end-pointing is given. Speech pauses are identified by a VAD based on energy and fundamental frequency. States for silence or speech pauses are included in the Markov models of the speech recognizer to enable a refined speech segmentation. An additional speaker change detection is omitted.

Automated speaker identification and speech recognition are now briefly summarized:

- **Front-end.** Standard speech enhancement techniques are applied to the speech signal to limit the influence of the acoustic environment. Feature vectors are extracted from the enhanced speech signal to be used for speech recognition and speaker identification. The computational complexity of two separate algorithms is avoided. The speech recognizer dominates the choice of feature vectors because speaker identification is only intended to support speech recognition as an optional component of the complete system. The basic architecture of the speech recognizer is preserved. For each speaker the long-term energy and mean normalization is continuously adapted. Since speaker changes are assumed to be rare, users can be expected to speak several utterances. Therefore, only the simple solution of one active feature extraction is considered. When a speaker turn

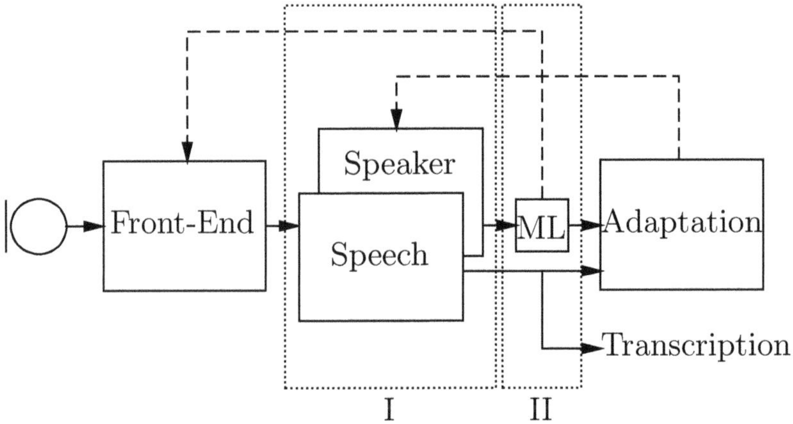

Fig. 5.5 System architecture for joint speaker identification and speech recognition. One front-end is employed for speaker specific feature extraction. Speaker specific codebooks (I) are used to decode the spoken phrase and to estimate the speaker identity (II) in a single step. Both results are used for speaker adaptation to enhance future speaker identification and speech recognition. Figure is taken from [Herbig et al., 2010e].

occurs, the front-end discards the mean and energy modifications originating from the last utterance. The parameter set which belongs to the identified speaker is used to process the next utterance.

- **Speech recognition.** Appropriate speaker specific codebooks are selected by a local speaker identification to achieve an optimal decoding of the recorded utterance. The speech recognizer can react quickly to speaker changes. However, codebook selection is expected to be less confident compared to speaker identification on an utterance-level. Speech recognition delivers a transcription of the spoken phrases on an utterance-level. The state alignment of the current utterance can be used in the subsequent speaker adaptation to optimize the corresponding codebook.
- **Speaker identification.** The likelihood values $p(\mathbf{x}_t|\Theta_i)$, $i = 0, \ldots, N_{\mathrm{Sp}}$ of each time instance are buffered and the speaker identity is estimated on an utterance-level. The speaker identity is used for codebook adaptation and for speaker specific feature extraction.
- **Speaker adaptation.** The codebooks of the speech recognizer are initialized and adjusted based on the speaker identity and a transcription of the spoken phrase.

For each recorded utterance the procedure above is applied. This results in an enhanced recognition rate for speech recognition and speaker identification as well.

If the computational load is not a critical issue, a set of parallel speaker specific feature extractions may be realized to provide an optimal feature normalization for each codebook. Alternatively, the complete utterance may be re-processed when a speaker change is detected. However, this introduces an additional latency and overhead. The performance decrease which is caused by only one active feature normalization will be evaluated in the experiments.

5.3 Reference Implementation

In the preceding sections a unified approach for speaker identification and speech recognition was introduced. Alternatively, standard techniques for speaker identification and speaker adaptation can be employed. In Sect. 2.4.2 a standard technique for speaker identification based on GMMs purely optimized to capture speaker characteristics was described. In Sect. 2.6 standard methods for speaker adaptation were introduced to initialize and continuously adjust speaker specific GMMs.

This speaker identification technique is used here without further modifications as a reference for the unified approach discussed. Combined with the speaker specific speech recognizer of Sect. 5.2.1 a reference implementation can be easily obtained. It is intended to deliver insight into the speech recognition and speaker identification rates which can be achieved by such an approach.

First, the specifications and the implementation of the speaker identification are given. Then the system architecture of the alternative system is described. The experiments of the preceding chapter will be discussed with respect to this reference.

5.3.1 Speaker Identification

Speaker identification is subsequently realized by common GMMs purely representing speaker characteristics. Fig. 5.6 shows the corresponding block diagram and has to be viewed as a detail of Fig. 5.7 specifying speaker identification and front-end.

A speaker independent UBM is used as template for speaker specific GMMs. Several UBMs comprising $32, 64, 128$ or 256 Gaussian distributions with diagonal covariance matrices have been examined in the experiments. The feature vectors comprise 11 mean normalized MFCCs and delta features. The 0 th cepstral coefficient is replaced by a normalized logarithmic energy estimate.

The UBM was trained by the EM algorithm. About 3.5 h speech data originating from 41 female and 36 male speakers of the USKCP[1] database was incorporated into the UBM training. For each speaker about 100 command

[1] The USKCP is an internal speech database for in-car applications. The USKCP was collected by TEMIC Speech Dialog Systems, Ulm, Germany.

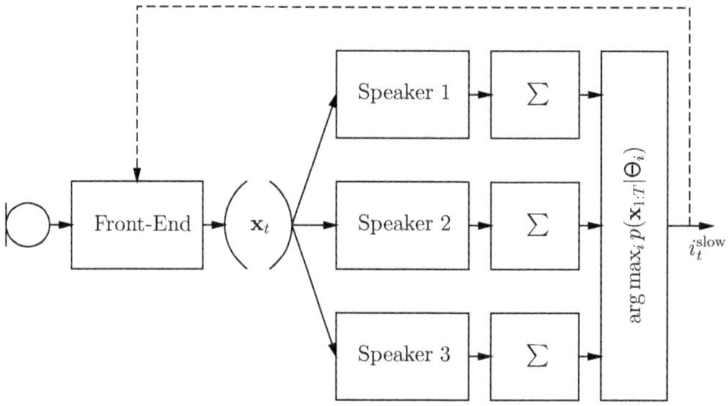

Fig. 5.6 Block diagram of speaker identification exemplified for 3 speakers. Each speaker is represented by one GMM. The accumulated log-likelihood is calculated based on the iid assumption. The speaker model with the highest likelihood is selected on an utterance-level. The speaker identity is used for speaker adaptation. Additionally, speaker specific feature normalization is controlled by speaker identification.

and control utterances such as navigation commands, spelling and digit loops are available. The language is US-English. The training was performed only once and the resulting UBM has been used without any further modifications in the experiments.

When a new speaker is enrolled, a new GMM is initialized by MAP adaptation according to (2.85) based on the first two utterances. Each speaker model is continuously adapted as soon as an estimate of the speaker identity is available.

The log-likelihood $\log p\,(\mathbf{x}_{1:T_u}|\Theta_i)$ of each utterance $\mathbf{x}_{1:T_u}$ is iteratively computed for all speakers $i = 1, \ldots, N_{\text{Sp}}$ in parallel to speech recognition. Θ_i denotes the speaker specific parameter set. The iid assumption is applied to simplify likelihood computation by accumulating the log-likelihood values of each time instance as described by (2.46). The speaker with the highest likelihood score is identified as the current speaker

$$i_u^{\text{slow}} = \arg\max_i \left\{\log(p(\mathbf{x}_{1:T_u}|\Theta_i))\right\}. \tag{5.26}$$

To guarantee that always an appropriate codebook is employed for speech decoding, codebook selection is realized as discussed before. However, speaker adaptation is based on the result of this speaker identification.

5.3.2 System Architecture

The reference system is realized by a parallel computation of speaker identification with GMMs on an utterance-level and the speaker specific speech

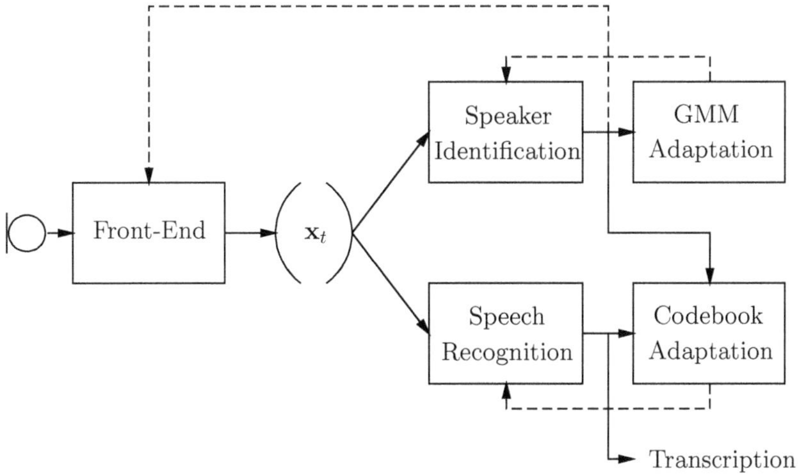

Fig. 5.7 System architecture for parallel speaker identification and speaker specific speech recognition. Codebook selection is implemented as discussed before. Speaker identification is realized by an additional GMM for each speaker. Figure is taken from [Herbig et al., 2010b].

recognition previously described. The setup is depicted in Fig. 5.7. The work flow of the parallel processing can be summarized as follows [Herbig et al., 2010b]:

- **Front-end.** The recorded speech signal is preprocessed to reduce background noises. Feature vectors are extracted as discussed before.
- **Speech recognition.** Appropriate speaker specific codebooks are selected for the decoding of the recorded utterance as presented in Sect. 5.2.
- **Speaker identification.** One GMM is trained for each speaker to represent speaker characteristics. A standard speaker identification technique is applied to estimate the speaker's identity on an utterance-level. The codebooks of the speech recognizer and the corresponding GMMs are adapted according to this estimate.
- **Speaker adaptation.** GMM models and the corresponding codebooks are continuously updated given the guess of the speaker identity. Codebook adaptation is implemented by the speaker adaptation scheme described in Sect. 4.2. GMMs are adjusted by MAP adaptation introduced in Sect. 2.6.3. Adaptation accuracy is supported here by the moderate complexity of the applied GMMs.

5.4 Evaluation

In the following both realizations for speaker specific speech recognition combined with speaker identification are discussed. First, the joint approach and then the reference implementation are considered for closed-set and open-set scenarios.

5.4.1 Evaluation of Joint Speaker Identification and Speech Recognition

In the following experiments the joint speaker identification and speech recognition is investigated. First, experiments are discussed which investigate speaker identification for closed sets. Then the detection of unknown speakers is examined.

Closed-Set Speaker Identification

First, the speaker identification accuracy of the joint approach is examined based on a supervised experiment characterized by optimum conditions. Then a realistic scenario for a limited group of enrolled speakers is discussed. Finally, the influence of feature normalization on the performance of the system is considered.

Speaker Identification Accuracy

In a first step the performance of speaker identification is determined for the approach shown in Fig. 5.5. The correct speaker identity is used here for speaker adaptation to investigate the identification accuracy depending on the training level of speaker specific codebooks. Even though error propagation is neglected, one can get a first overview of the identification accuracy of the joint approach discussed in the preceding sections.

The test setup is structured as described in Fig. 4.1(a). It comprises 4 sets each with 50 speakers in a randomized order. For each speaker 200 utterances are used randomly. The likelihood scores of all speakers are recorded. The codebook of the target speaker is modified by speaker adaptation after each utterance as described in Chapter 4.

For evaluation, several permutations are used to randomly build speaker pools of a predetermined size. The resulting speaker identification rate is plotted in Fig. 5.8. For $\lambda = 4$ the best performance of speaker identification is achieved whereas accuracy decreases for higher values of λ. However, when λ tends towards zero, the identification rate drops significantly since unreliable ML estimates are used, especially during the initialization. The EV approach obviously loses for larger speaker sets. Since each speaker is represented by only 10 parameters, modeling accuracy is limited and the probability of false classifications rises in the long run compared to the combination of EV and ML estimates.

In Fig. 5.9 the result is depicted in more detail. It is represented by four evaluation bands based on I $= [1; 50[$, II $= [50; 100[$, III $= [100; 150[$ and

IV = [150; 200] utterances. The evolution of speaker specific codebooks versus their adaptation level is shown. Each curve represents a predetermined speaker pool size. According to Fig 5.9 it can be stated that the identification accuracy rises with improved speaker adaptation as expected. The identification rate is a function of the tuning parameter λ, the number of enrolled speakers N_{Sp} and training. Having analyzed the measured identification rates, a small parameter λ seems to be advantageous for an accurate speaker identification. Even in the last evaluation band a significant difference between $\lambda = 4$, $\lambda = 12$ and $\lambda = 20$ can be observed. The threshold λ has still an influence on adaptation accuracy even though 150 utterances have been accumulated. However, speaker identification is generally affected by a higher number of enrolled speakers since the probability of confusions increases.

In summary, Fig. 5.9 shows a consistent improvement of the identification rate for all setups as soon as sufficient adaptation data is accessible. The implementation with $\lambda = 4$ clearly outperforms the remaining realizations, especially for larger speaker sets. However, the enrollment phase which is characterized by the lowest identification rate is critical for long-term stability as shown subsequently.

Unsupervised Closed-Set Identification of 5 *enrolled Speakers*

In the following the closed-set scenario of a self-learning speaker identification combined with speech recognition and unsupervised speaker adaptation is investigated [Herbig et al., 2010d]. The subsequent considerations continue the discussion of the example in Sect. 4.3.3.

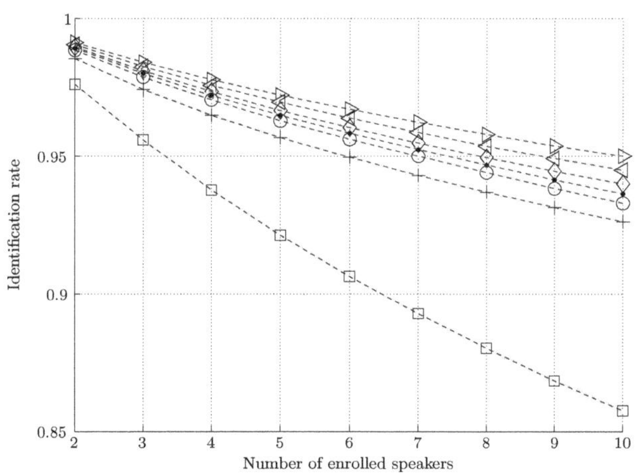

Fig. 5.8 Performance of the **self-learning speaker identification**. Speaker adaptation is supervised so that **no maladaptation** occurs. Several speaker pools are investigated for different parameters of speaker adaptation - ML ($\lambda \approx 0$) (+), $\lambda = 4$ ($\triangleright$), $\lambda = 8$ ($\triangleleft$), $\lambda = 12$ ($\diamond$), $\lambda = 16$ ($\cdot$), $\lambda = 20$ ($\circ$) and EV ($\lambda \rightarrow \infty$) ($\square$).

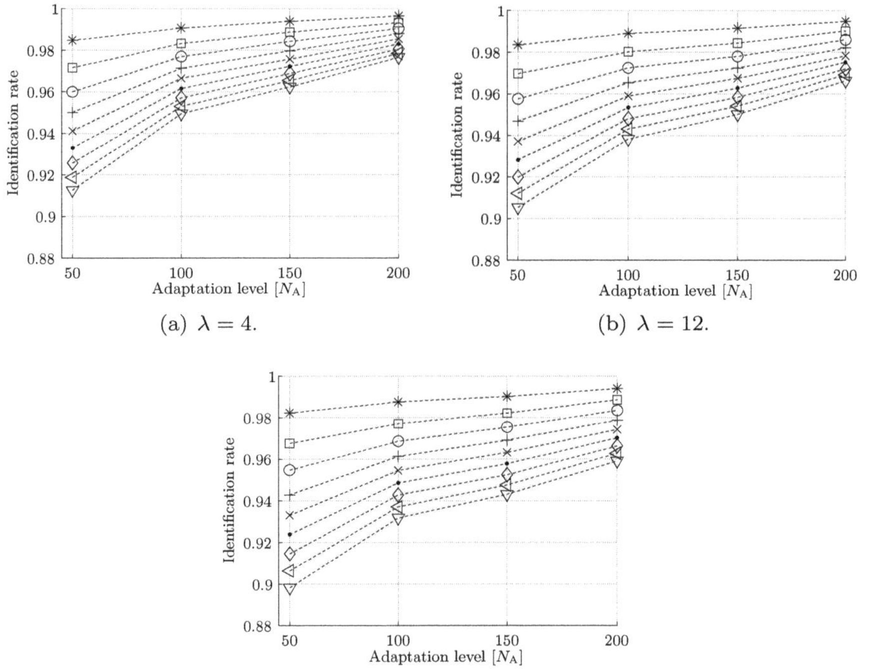

(a) $\lambda = 4$.

(b) $\lambda = 12$.

(c) $\lambda = 20$.

Fig. 5.9 Performance of the self-learning speaker identification versus adaptation stage. N_A denotes the number of utterances used for speaker adaptation. Several speaker pools are considered - 2 ($\star$), 3 ($\square$), 4 ($\circ$), 5 ($+$), 6 ($\times$), 7 ($\cdot$), 8 ($\diamond$), 9 ($\triangleleft$) and 10 ($\triangledown$) enrolled speakers listed in top-down order.

Closed-set speaker identification was evaluated on 60 sets. Each set comprises 5 enrolled speakers. In Sect. 4.3.3 the speaker identity was given for speaker adaptation leading to optimally trained speaker profiles of the speech recognizer.

The goal is now to determine what speech recognition and identification results can be achieved if this knowledge about the true speaker *identity* (ID) is not given. Only the first two utterances of a new speaker are indicated and then the current speaker has to be identified in a completely unsupervised manner. The experiment described in Sect. 4.3.3 has been repeated with unsupervised speaker identification. The resulting speech recognition and speaker identification rates are summarized in Table 5.1. In addition, the percentage of the utterances which are rejected by the speech recognizer is given.

The results show significant improvements of the WA with respect to the baseline and short-term adaptation. The two special cases ML ($\lambda \approx 0$) and EV ($\lambda \to \infty$) clearly fall behind the combination of both adaptation

Table 5.1 Comparison of the different adaptation techniques for **self-learning speaker identification**. Speaker pools with **5 enrolled speakers** are considered. Table is taken from [Herbig et al., 2010b].

Speaker adaptation	Rejected [%]	WA [%]	Speaker ID [%]
Baseline	-	85.23	-
Short-term adaptation	-	86.13	-
EV-MAP adaptation			
ML ($\lambda \approx 0$)	2.10	86.89	81.54
$\lambda = 4$	2.09	88.10	**94.64**
$\lambda = 8$	2.11	88.17	93.49
$\lambda = 12$	2.10	88.16	92.42
$\lambda = 16$	2.09	88.18	92.26
$\lambda = 20$	2.06	**88.20**	91.68
EV ($\lambda \to \infty$)	2.11	87.51	84.71
MAP adaptation			
$\eta = 4$	2.02	87.47	87.43
Typical errors			
min		± 0.23	± 0.16
max		± 0.25	± 0.28

Table 5.2 Comparison of different adaptation techniques for **self-learning speaker identification**. Speaker pools with **5 enrolled speakers** are considered. Speaker identification and speech recognition results are given for several evaluation bands - I = $[1; 50[$, II = $[50; 100[$, III = $[100; 250]$ utterances.

Speaker adaptation	I		II		III	
	WA [%]	ID [%]	WA [%]	ID [%]	WA [%]	ID [%]
Baseline	84.51	-	86.84	-	85.02	-
Short-term adaptation	85.89	-	87.22	-	85.87	-
EV-MAP adaptation						
ML ($\lambda \approx 0$)	85.49	78.79	87.45	79.51	87.35	83.13
$\lambda = 4$	87.42	**92.61**	89.18	**93.93**	88.04	**95.54**
$\lambda = 8$	87.46	90.83	**89.33**	92.44	88.11	94.71
$\lambda = 12$	87.51	89.94	**89.33**	91.69	88.07	93.48
$\lambda = 16$	87.56	89.77	89.31	91.40	88.09	93.37
$\lambda = 20$	**87.57**	89.38	89.22	90.99	**88.15**	92.67
EV ($\lambda \to \infty$)	87.38	85.08	88.56	84.75	87.21	84.57
Typical errors						
min	± 0.53	± 0.43	± 0.49	± 0.38	± 0.30	± 0.19
max	± 0.58	± 0.67	± 0.54	± 0.65	± 0.33	± 0.35

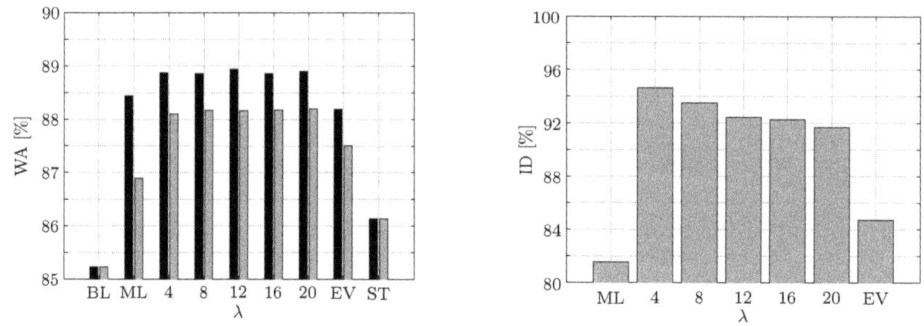

(a) Results for speech recognition. (b) Results for speaker identification.

Fig. 5.10 Comparison of supervised speaker adaptation with predefined speaker identity (black) and the joint speaker identification and speech recognition (dark gray) described in this chapter. The speaker independent baseline (BL) and short-term adaptation (ST) are depicted for comparison. Figure is taken from [Herbig et al., 2010d].

techniques. Furthermore, no eminent difference in WA can be observed for $4 \leq \lambda \leq 20$ [Herbig et al., 2010d].

It becomes obvious that speaker identification can be optimized independently from the speech recognizer and seems to reach an optimum of 94.64 % for $\lambda = 4$. This finding also agrees with the considerations concerning the identification accuracy sketched above. For higher values the identification rates drop significantly. Obviously, speaker characteristics can be captured despite limited adaptation data and the risk of error propagation. A comparison of the evaluation bands shows a consistent behavior for all realizations.

In Fig. 5.10 the speech recognition rates achieved by the joint speaker identification and speech recognition are graphically compared with the supervised experiment in Sect. 4.3. Additionally, the speaker identification rates are depicted. It can be stated that this first implementation is already close to the upper bound given by the supervised experiment and is significantly better than the baseline. Again, no eminent differences in the WA can be stated but a significant decrease in the speaker identification rate can be observed for increasing λ.

Potential for Further Improvements

Finally, the influence of speaker specific feature extraction on speaker identification and speech recognition is quantified for the scenario discussed before. The experiment has been repeated with supervised energy and mean normalization. Before an utterance is processed, the correct parameter set is loaded and continuously adapted. Speaker identification, speech recognition and speaker adaptation remain unsupervised.

Table 5.3 Results for joint speaker identification and speech recognition with **supervised speaker specific feature extraction**. The feature vectors are normalized based on the parameter set of the target speaker. The parameters are continuously adapted.

Speaker adaptation	Rejected [%]	WA [%]	Speaker ID [%]
Baseline	-	85.23	-
Short-term adaptation	-	86.13	-
EV-MAP adaptation			
$\lambda = 4$	2.05	88.47	**97.33**
$\lambda = 8$	2.11	88.55	96.66
$\lambda = 12$	2.06	**88.60**	96.22
$\lambda = 16$	2.09	88.56	95.88
$\lambda = 20$	2.08	**88.61**	95.75
Typical errors			
min		± 0.23	± 0.12
max		± 0.25	± 0.15

When Table 5.3 is compared with Table 5.1, it becomes evident that feature normalization influences speaker identification significantly. Furthermore, WA can be increased when the feature vectors are accurately normalized. Even though this is an optimal scenario concerning feature extraction, a clear improvement for speaker identification can be expected when feature extraction is operated in parallel.

Open-Set Speaker Identification

In the preceding section speaker identification was evaluated for a closed-set scenario. To achieve a speech controlled system which can be operated in a completely unsupervised manner, unknown speakers have to be automatically detected. This enables the initialization of new codebooks which can be adapted to new users.

In Fig. 5.11 an open-set scenario is considered [Herbig et al., 2010a]. Again, speaker specific codebooks are evaluated as discussed before. Unknown speakers are detected by the threshold decision in (5.25). If this threshold is not exceeded by any speaker specific codebook, an unknown speaker is hypothesized. Speaker adaptation is supervised so that maladaptation with respect to the speaker identity is neglected.

The performance of the in-set / out-of-set classification is evaluated by the so-called *Receiver Operator Characteristics* (ROC) described in Sect. A.3. The detection rate of unknown speakers is plotted versus false alarm rate.

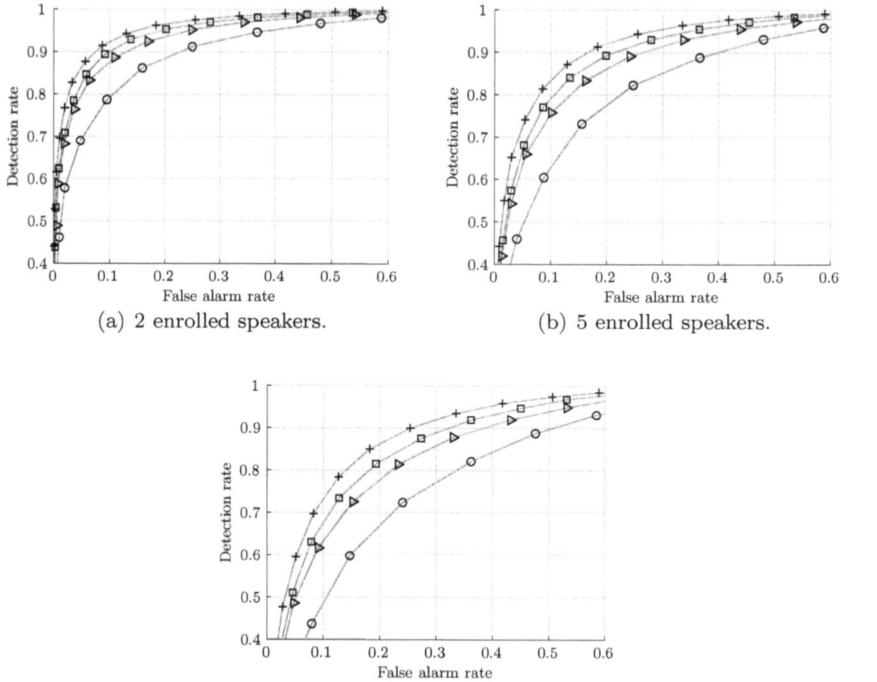

(a) 2 enrolled speakers.

(b) 5 enrolled speakers.

(c) 10 enrolled speakers.

Fig. 5.11 Detection of unknown speakers based on log-likelihood ratios of the speaker specific codebooks and standard codebook. Speaker adaptation ($\lambda = 4$) is supervised. Maladaptation with respect to the speaker identity is neglected. Several evaluation bands are shown - $N_A \leq 20$ ($\circ$), $20 < N_A \leq 50$ ($\triangleright$), $50 < N_A \leq 100$ ($\square$) and $100 < N_A \leq 200$ ($+$). Confidence intervals are given by a gray shading. Figure is taken from [Herbig et al., 2010a].

The ROC curves in Fig. 5.11 show that open-set speaker identification is difficult when speaker models are trained on only a few utterances. For $N_A \leq 20$ both error rates are unacceptably high even if only two enrolled speakers are considered. At least 50 utterances are necessary for adaptation to achieve a false alarm and miss rate less than 10 %. For 5 or 10 enrolled speakers the detection rate of unknown speakers is even worse.

In summary, unknown speakers can be detected by applying a simple threshold to the log-likelihood ratios of the speaker specific codebooks and the standard codebook. However, the detection rates achieved in the experiments show that it is difficult for such a system to detect new users based on only one utterance as expected. A global threshold seems to be inappropriate since the training of each speaker model should be taken into consideration [Herbig et al., 2010a].

In a practical implementation this problem would be even more complicated when mixed codebooks are used or when several codebooks belong to one speaker. It is hard to imagine that a reasonable implementation of an unsupervised speech controlled system can be realized with these detection rates, especially when 10 speakers are enrolled. Thus, a more robust technique is required for a truly unsupervised out-of-set detection.

5.4.2 Evaluation of the Reference Implementation

In the following the performance of the reference implementation is examined. Several settings of the reference implementation are compared for the closed-set and open-set identification task [Herbig et al., 2010a,b].

Closed-Set Speaker Identification

First, the speaker identification accuracy of the reference implementation is examined based on a supervised experiment characterized by optimum conditions. Then a realistic closed-set scenario is considered for a limited group of enrolled speakers.

Speaker Identification Accuracy

The performance of speaker identification is determined for the approach shown in Fig. 5.7. The correct speaker identity is used here for speaker adaptation to describe the identification accuracy of speaker specific GMMs. Speaker identification and GMM training are realized as described in Sect. 5.3. If not indicated otherwise, mean vector adaptation is performed to enhance speaker specific GMMs.

In Fig. 5.12 the averaged accuracy is compared for dedicated realizations in analogy to Fig. 5.8. GMMs containing a high number of Gaussian distributions show the best performance probably because a more individual adjustment is possible. A comparison with Fig. 5.8 reveals that the codebooks of the speech recognizer cannot be expected to model speaker characteristics as accurately as GMMs purely optimized for speaker identification. However, in terms of performance all realizations yield good results. It should be emphasized that this consideration is based on a supervised adaptation. Maladaptation due to speaker identification errors is neglected. The further experiments will show the deficiencies of the reference implementation to accurately track the current speaker.

Unsupervised Speaker Identification of 5 *enrolled Speakers*

A realistic closed-set application for self-learning speaker identification combined with speech recognition and unsupervised speaker adaptation is investigated. The subsequent considerations continue the discussion of the example in Sect. 4.3.3 and Sect. 5.4.1. Closed-set speaker identification is evaluated on 60 sets. Each set comprises 5 enrolled speakers.

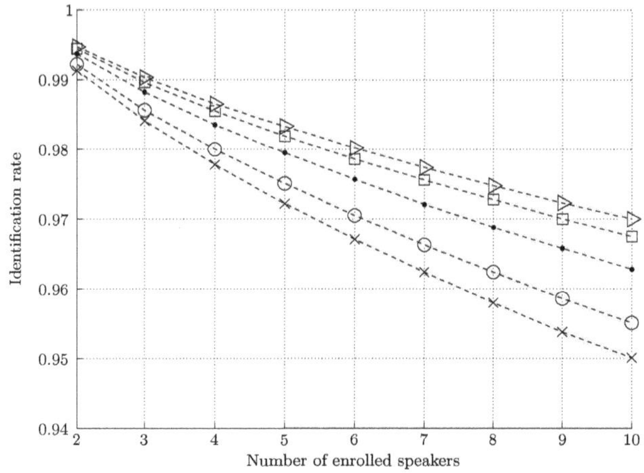

Fig. 5.12 Performance of the **reference implementation**. Speaker identification is implemented by GMMs comprising 32 (∘), 64 (·), 128 (□) or 256 (▹) Gaussian distributions. MAP adaptation ($\eta = 4$) is only applied to mean vectors. For comparison, the codebooks ($\times$) of the speech recognizer are adjusted using $\lambda = 4$. For all implementations **maladaptation** is **not** considered. Confidence intervals are given by a gray shading.

In Table 5.4 the results of this scenario are presented for several implementations with respect to the number of Gaussian distributions and values of parameter η. Both the speaker identification and speech recognition rate reach an optimum for $\eta = 4$ and $N = 64$ or 128. For higher values of η this optimum is shifted towards a lower number of Gaussian distributions as expected. Since the learning rate of the adaptation algorithm is reduced, only a reduced number of distributions can be efficiently estimated at the

Table 5.4 Realization of **parallel speaker identification and speech recognition**. Speaker identification is implemented by several GMMs comprising 32, 64, 128 and 256 Gaussian distributions. **MAP adaptation of mean vectors** is used. Codebook adaptation uses $\lambda = 12$. Table is taken from [Herbig et al., 2010b].

MAP	$\eta = 4$		$\eta = 8$		$\eta = 12$		$\eta = 20$	
N	WA [%]	ID [%]	WA [%]	ID [%]	WA [%]	ID [%]	WA [%]	ID [%]
32	88.01	88.64	**88.06**	88.17	**87.98**	87.29	**87.97**	**87.50**
64	**88.13**	91.09	**88.06**	**89.64**	**87.98**	**87.92**	87.92	85.30
128	88.04	**91.18**	87.94	87.68	87.87	84.97	87.82	80.09
256	87.92	87.96	87.97	85.59	87.90	81.20	87.73	76.48
Typical errors								
min	±0.23	±0.21	±0.23	±0.22	±0.23	±0.24	±0.23	±0.24
max	±0.23	±0.24	±0.23	±0.25	±0.23	±0.28	±0.23	±0.31

Table 5.5 Detailed investigation of the WA and identification rate on several evaluation bands. **MAP adaptation of mean vectors** is used. Codebook adaptation uses $\lambda = 12$. Several evaluation bands are considered - I = [1; 50[, II = [50; 100[, III = [100; 250] utterances.

MAP adaptation $\eta = 4$	I WA [%]	I ID [%]	II WA [%]	II ID [%]	III WA [%]	III ID [%]
Baseline	84.51	-	86.84	-	85.02	-
Short-term adaptation	85.89	-	87.22	-	85.87	-
Number of Gaussian distributions N						
32	87.41	86.14	**89.17**	88.65	87.89	89.46
64	**87.49**	**88.76**	89.13	**91.13**	**88.09**	91.84
128	87.38	88.73	89.03	90.80	88.02	**92.12**
256	87.31	85.53	88.86	87.39	87.88	88.95
Typical errors						
min	±0.53	±0.52	±0.50	±0.45	±0.30	±0.25
max	±0.58	±0.57	±0.54	±0.53	±0.33	±0.29

Table 5.6 Realization of **parallel speaker identification and speech recognition**. Speaker identification is implemented by several GMMs comprising 32, 64, 128 or 256 Gaussian distributions. **MAP adaptation of weights and mean vectors** is used. Codebook adaptation uses $\lambda = 12$. Table is taken from [Herbig et al., 2010b].

MAP N	$\eta = 4$ WA [%]	$\eta = 4$ ID [%]	$\eta = 8$ WA [%]	$\eta = 8$ ID [%]	$\eta = 12$ WA [%]	$\eta = 12$ ID [%]	$\eta = 20$ WA [%]	$\eta = 20$ ID [%]
32	87.92	87.24	87.97	88.24	87.97	87.61	88.02	**87.04**
64	**88.11**	90.59	**88.06**	**89.99**	**88.03**	**88.80**	**87.93**	86.64
128	**88.11**	91.32	88.03	89.42	**88.03**	88.10	87.91	84.26
256	**88.10**	**91.62**	87.97	88.71	**88.02**	86.01	87.88	82.88
Typical errors								
min	±0.23	±0.20	±0.23	±0.22	±0.23	±0.23	±0.23	±0.24
max	±0.23	±0.24	±0.23	±0.23	±0.23	±0.25	±0.23	±0.27

beginning. The performance of the speech recognizer is marginally reduced with higher η [Herbig et al., 2010b].

Table 5.5 contains the corresponding detailed results of speaker identification and speech recognition obtained on different evaluation bands for $\eta = 4$. It can be concluded that 32 Gaussian distributions are not sufficient whereas more than 128 Gaussian distributions seems to be oversized. The highest identification rates of 91.18 % and 91.09 % have been achieved with 64 and 128 distributions. Comparing Table 5.2 and Table 5.5 similar results for speech recognition can be observed.

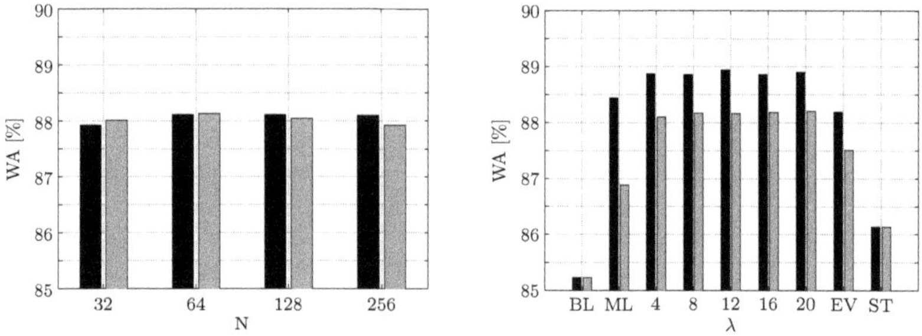

(a) Speech recognition rates of the reference implementation. MAP adaptation ($\eta = 4$) of mean vectors and weights (black) and only mean vectors (dark gray) are depicted.

(b) Speech recognition rates of the unified approach. Results are shown for speaker adaptation with predefined speaker identity (black) as well as for joint speaker identification and speech recognition (dark gray). The speaker independent baseline (BL) and short-term adaptation (ST) are shown for comparison.

Fig. 5.13 Comparison of the **reference implementation** (left) and the **joint speaker identification and speech recognition** (right) with respect to **speech recognition**. Figure is taken from [Herbig et al., 2010b].

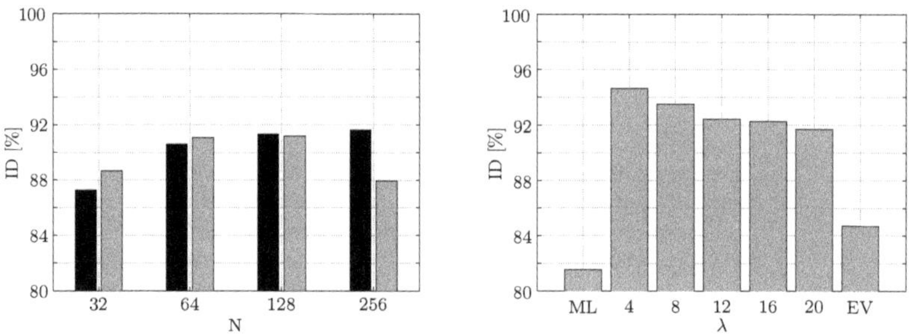

(a) Speaker identification rates of the reference implementation. MAP adaptation ($\eta = 4$) of mean vectors and weights (black) and only mean vectors (dark gray) are depicted.

(b) Speaker identification rates of the joint speaker identification and speech recognition.

Fig. 5.14 Comparison of the **reference implementation** (left) and the **joint speaker identification and speech recognition** (right) with respect to **speaker identification**. Figure is taken from [Herbig et al., 2010b].

For the next experiment not only mean vectors but also weights are modified. The results are summarized in Table 5.6. In the preceding experiment the speaker identification accuracy could be improved for $\eta = 4$ by increasing the number of Gaussian distributions to $N = 128$. For $N = 256$ the identification rate dropped significantly. Now a steady improvement and an optimum

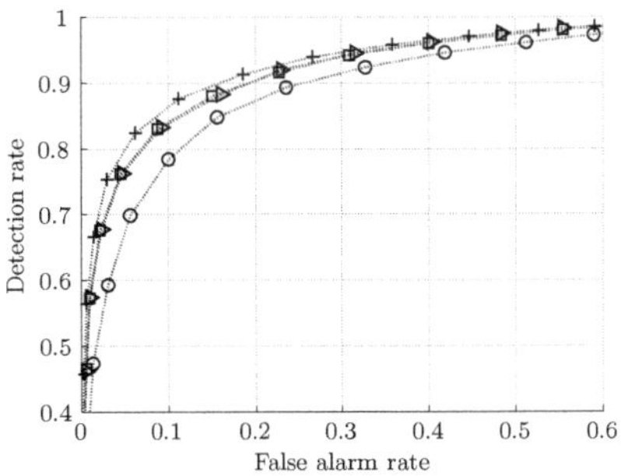

(a) 2 enrolled speakers.

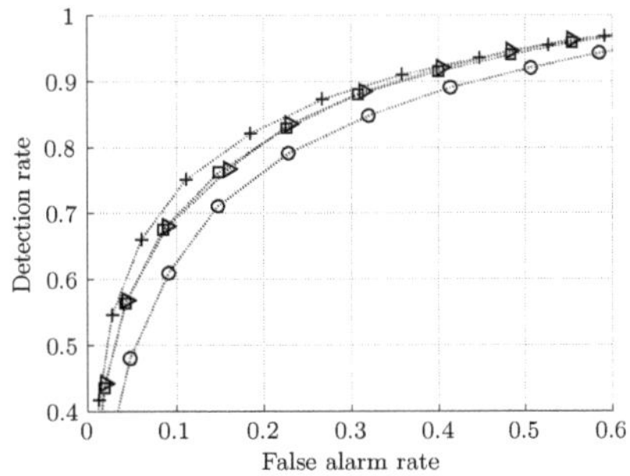

(b) 5 enrolled speakers.

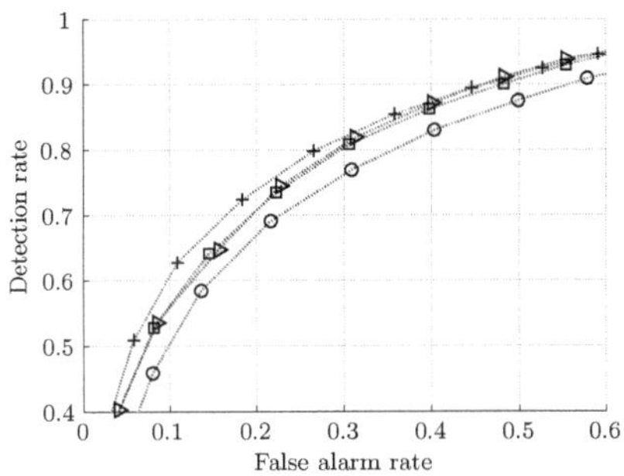

(c) 10 enrolled speakers.

Fig. 5.15 Detection of unknown speakers based on log-likelihood ratios of the speaker specific GMMs and UBM. Speaker identification is realized by GMMs with **64 Gaussian distributions**. $\eta = 4$ is used to adapt the mean vectors. Maladaptation is neglected. Several evaluation bands are investigated - $N_{\mathrm{A}} \leq 20$ ($\circ$), $20 < N_{\mathrm{A}} \leq 50$ ($\triangleright$), $50 < N_{\mathrm{A}} \leq 100$ ($\square$) and $100 < N_{\mathrm{A}} \leq 200$ ($+$). Confidence intervals are given by a gray shading.

of 91.62 % can be observed for $N = 256$. However, the identification rate approaches a limit. For $\eta = 4$ doubling the number of Gaussian distributions from 32 to 64 results in 26 % relative error rate reduction. The relative error rate reduction which is achieved by increasing the number of Gaussian distributions from 128 to 256 is about 3.5 %. The optimum for speech recognition is again about 88.1 % WA [Herbig et al., 2010b].

Finally, the results of the unified approach characterized by an integrated speaker identification and the reference implementation are compared in Fig. 5.13 and Fig. 5.14. Both mean vector and weight adaptation are depicted for $\eta = 4$ representing the best speech recognition and speaker identification rates obtained by GMMs [Herbig et al., 2010b].

In summary, similar results for speech recognition are achieved but identification rates are significantly worse compared to the experiments discussed before. This observation supports again the finding that the speech recognition accuracy is relatively insensitive with respect to moderate error rates

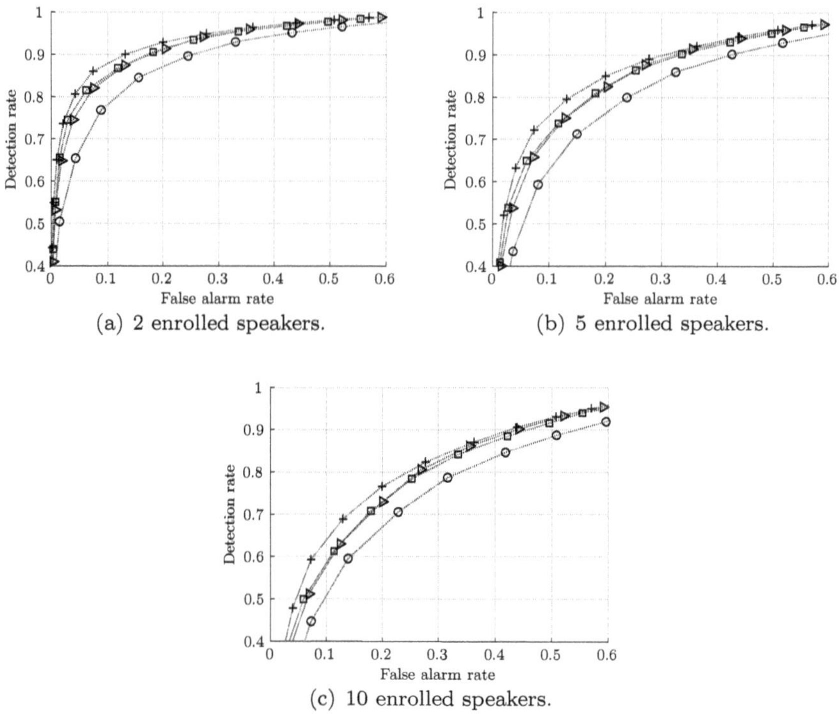

Fig. 5.16 Detection of unknown speakers based on log-likelihood ratios of the speaker specific GMMs and UBM. Speaker identification is realized by GMMs with **256 Gaussian distributions.** $\eta = 4$ is used for adapting the mean vectors. Maladaptation is neglected. Several evaluation bands are investigated - $N_A \leq 20$ (○), $20 < N_A \leq 50$ (▷), $50 < N_A \leq 100$ (□) and $100 < N_A \leq 200$ (+). Confidence intervals are given by a gray shading. Figure is taken from [Herbig et al., 2010a].

of speaker identification. Thus, different strategies can be applied to identify speakers without affecting the performance of the speech recognizer as long as a robust codebook selection is employed in speech decoding.

Open-Set Speaker Identification

In the following the open-set scenario depicted in Fig. 5.11 is examined for GMMs purely optimized to identify speakers [Herbig et al., 2010a]. Unknown speakers are detected by a threshold decision based on log-likelihood ratios of speaker specific GMMs and UBM similar to (2.49). The log-likelihood scores are normalized by the length of the utterance. The corresponding ROC curves are depicted in Fig. 5.15.

The ROC curves show again the limitations of open-set speaker identification to detect unknown speakers, especially when speaker models are trained

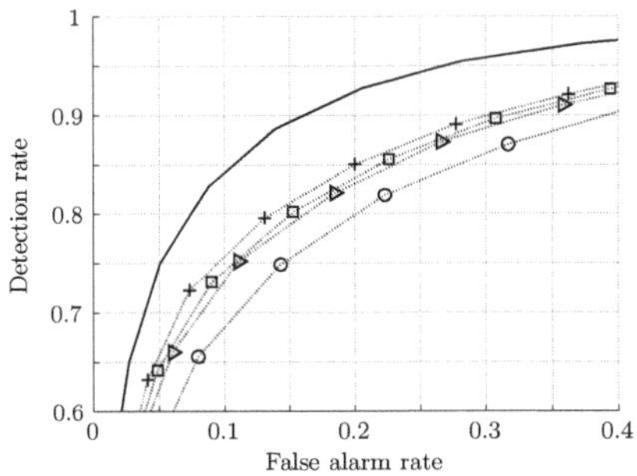

Fig. 5.17 Detection of unknown speakers. Speaker specific codebooks (solid line) are compared to GMMs comprising 32 (∘), 64 (▷), 128 (□) and 256 (+) Gaussian densities for $100 < N_{\mathrm{A}} \leq 200$. MAP adaptation ($\eta = 4$) is employed to adjust the mean vectors. Figure is taken from [Herbig et al., 2010a].

on only a few utterances. Compared to Fig. 5.11 the detection rates are significantly worse. It becomes obvious that these error rates are unacceptably high for a direct implementation in an unsupervised complete system.

In Fig. 5.16 the same experiment is repeated with GMMs comprising 256 Gaussian distributions. In contrast to the former implementation, the performance is clearly improved. However, the joint approach of speaker identification and speech recognition still yields significantly better results.

To compare the influence of the number of Gaussian densities on the detection accuracy, all reference implementations and a specific realization with speaker specific codebooks are shown in Fig. 5.17. Only the case of extensively trained speaker models ($100 < N_{\mathrm{A}} < 200$) is examined. When the number of Gaussian distributions is increased from 32 to 64, a clear improvement can be observed. However, the detection accuracy approaches a limit when a higher number of Gaussian distributions is used. It becomes obvious that the accuracy of all reference implementations examined here is inferior to the codebook based approach.

5.5 Summary and Discussion

Two approaches have been developed to solve the problem of an unsupervised system comprising self-learning speaker identification and speaker specific speech recognition.

First, a unified approach for simultaneous speaker identification and speech recognition was described. Then an alternative system comprising a standard technique for speaker identification was introduced.

Both systems have several advantages and drawbacks which should be discussed. Both implementations do not modify the basic architecture of the speech recognizer. Speaker identification and speech recognition use an identical front-end so that a parallel feature extraction for speech recognition and speaker identification is avoided. Computation on different time scales was introduced. A trade-off between a fast but less accurate speaker identification for speech recognition and a delayed but more confident speaker identification for speaker adaptation has been developed: Speaker specific speech recognition is realized by an on-line codebook selection. On an utterance level the speaker identity is estimated in parallel to speech recognition. Multiple recognitions are not required. Identifying the current user enables a speech recognizer to create and continuously adapt speaker specific codebooks. This enables a higher recognition accuracy in the long run.

94.64 % speaker identification rate and 88.20 % WA were achieved by the unified approach for $\lambda = 4$ and $\lambda = 20$, respectively. The results for the baseline and the corresponding upper bound were 85.23 % and 88.90 % WA according to Table 4.1 [Herbig et al., 2010b].

For the reference system several GMMs are required for speaker identification in addition to the HMMs of the speech recognizer. Complexity therefore increases since both models have to be evaluated and adapted. Under perfect conditions higher speaker identification accuracies could be achieved with the reference implementation in the experiments carried out. Under realistic conditions a speaker identification rate of 91.18 % was achieved for 128 Gaussian distributions and $\eta = 4$ when only the mean vectors were adapted. The best speech recognition result of 88.13 % WA was obtained for 64 Gaussian distributions. By adapting both the mean vectors and weights, the speaker identification rate could be increased to 91.62 % for 256 Gaussian distributions and $\eta = 4$. The WA remained at the same level [Herbig et al., 2010b].

However, the detection rates of unknown speakers were significantly worse compared to the unified approach. Especially for an unsupervised system this out-of-set detection is essential to guarantee long-term stability and to enhance speech recognition. Thus, only the unified approach is considered in the following.

The major drawback of both implementations becomes evident by the problem of unknown speakers which is not satisfactorily solved. This seems to be the most challenging task of a completely unsupervised speech recognizer, especially for a command and control application in an adverse environment. Therefore, further effort is necessary to extend the system presented in this chapter.

Another important issue is the problem of weakly, moderately and extensively trained speaker models. Especially during the initialization on the first few utterances of a new speaker, this approach bears the risk of severe error propagation. It seems to be difficult for weakly adapted speaker models to compete with well trained speaker models since adaptation always tracks speaker characteristics and residual influences of the acoustic environment. As discussed so far, both algorithms do not appropriately account for the training status of each speaker model except for codebook and GMM adaptation. The evolution of the system is not appropriately handled. Likelihood is expected to be biased.

Furthermore, both realizations do not provide any confidence measure concerning the estimated speaker identity. Utterances with uncertain origin cannot be rejected. Long-term stability might be affected because of maladaptation.

As discussed so far, the integration of additional information sources concerning speaker changes or identity are not addressed. For example, a beamformer, the elapsed time between two utterances or the completion of speech dialog steps may contribute to a more robust speaker change detection.

In summary, the joint approach depicted in Fig. 5.5 constitutes a complete system characterized by a unified statistical modeling, moderate complexity and a mutual benefit between speech and speaker modeling. It has some deficiencies due to the discrete decision of the speaker identity on an utterance-level, lack of confidence measures and therefore makes the detection of unknown speakers more complicated.

In the next chapter an extension is introduced which alleviates the drawbacks described above. A solution suitable for open-set scenarios will be presented.

6

Evolution of an Adaptive Unsupervised Speech Controlled System

In the preceding chapter a first solution and a reference implementation for combined speaker identification and speaker specific speech recognition were presented and evaluated as complete systems. However, some deficiencies discussed in the preceding section motivate to extend this first solution.

Speaker identification as discussed so far has been treated on an utterance level based on a likelihood criterion. The discussion in this chapter goes beyond that. Long-term speaker identification is investigated to track different speakers across several utterances. The evolution of the self-adaptive system is taken into consideration. Together this results in a more confident guess of the current speaker's identity. Long-term stability and the detection of unknown speakers can be significantly improved.

The motivation and goals of the new approach are discussed. Posterior probabilities are derived which reflect both the likelihood and adaptation stage of each speaker model. Combined with long-term speaker tracking on a larger time scale a flexible and robust speaker identification scheme may be achieved. This technique can be completed by the detection of unknown speakers to obtain a strictly unsupervised solution. Finally, an overview of the advanced architecture of the speech controlled system is presented. The problem of speaker variability is solved on different time scales. The evaluation results are given at the end of this chapter.

6.1 Motivation

In Chapter 5 a unified approach was introduced and evaluated on speech data recorded in an automotive environment. This technique is based on a unified modeling of speech and speaker related information. Especially for embedded systems in adverse environments, such an approach seems to be advantageous since a compact representation and fast retrieval of speaker characteristics for enhanced speech recognition are favorable. The latter became obvious due to the evaluation results and the improvements achieved.

T. Herbig, F. Gerl, and W. Minker: Self-Learning Speaker Identification, SCT, pp. 115–143.
springerlink.com

However, unknown speakers appear to be a more challenging problem. As discussed so far, the integrated speaker identification does not enable a robust detection of unknown speakers. Obviously, it is very difficult to detect new users only on a single utterance.

In addition, the problem of speaker models at highly different training levels should be addressed in a self-learning speech controlled system, especially if it is operated in an unsupervised manner. Error propagation is expected since weakly trained speaker models are characterized by a higher error rate with respect to the in-set / out-of-set detection. Confidence measures, e.g. posterior probabilities, are expected to be more sophisticated than decisions based on likelihoods.

Finally, it may be advantageous to integrate additional devices, e.g. a beamformer, to support speaker tracking. Even though this aspect goes beyond the scope of this book, the following considerations can be easily extended as will be shown later.

Therefore, the solution presented in the preceding chapter shall be extended. The strengths of the unified approach shall be kept but its deficiencies shall be removed or at least significantly moderated. The problem of speaker tracking is now addressed on a long-term time scale. The goal is to buffer a limited number of utterances in an efficient manner. A path search algorithm is applied to find the optimal assignment to the enrolled speakers or an unknown speaker.

In addition, the training status of speaker specific codebooks is considered. As discussed before, the evaluation of each codebook in (5.23) does not distinguish between marginally and extensively adapted speaker models. Long-term speaker adaptation is characterized by a high number of parameters which allow a highly individual adjustment. This degree of individualism results in an improved match between statistical model and observed feature vectors. Thus, the likelihood is expected to converge to higher values for extensively adapted codebooks.

In a first step, a posterior probability is introduced based on a single utterance by taking this likelihood evolution into consideration. Then long-term speaker tracking is employed to extend this posterior probability to a series of successive utterances. A robust estimate of the speaker identity can be given by the MAP criterion. A confidence measure is automatically obtained since the information about the speaker's identity is kept as a probability instead of discrete ML estimates. Unknown speakers can be detected as a by-product of this technique.

6.2 Posterior Probability Depending on the Training Level

In this section the effect of different training or adaptation levels on likelihood values is investigated for a closed-set scenario. Newly initialized codebooks are less individually adapted compared to extensively trained speaker models. On average a smaller ratio of the speaker specific log-likelihoods $\mathcal{L}_u^i$ and the log-likelihood of the standard codebook $\mathcal{L}_u^0$ is expected compared to extensively trained models since all codebooks evolve from the standard codebook [Herbig et al., 2010c].

The goal is to employ posterior probabilities instead of likelihoods which integrate the log-likelihood scores $\mathcal{L}_u^i$ and the training status of each speaker model [Herbig et al., 2010e]. Thus, it is proposed to consider likelihood values as random variables with respect to the adaptation status and to learn the corresponding probability density function in a training. The latter has to be performed only once as described subsequently.

6.2.1 Statistical Modeling of the Likelihood Evolution

For a statistical evaluation of the likelihood distribution which is achieved in an unsupervised speech controlled system the scores of 100,000 utterances for target and non-target speakers were investigated in a closed-set scenario. In order to have a realistic profile of internal parameters, e.g. log-likelihood and training level, concerning speaker changes the speakers were organized in 20 sets containing 25 enrolled speakers. The composition was randomly chosen and needed not to be gender balanced. In total, 197 native speakers of the USKCP[1] development database were employed. For each speaker approximately 300 − 400 utterances were recorded under different driving and noise conditions. For each speaker a training set of 200 utterances was separated. Users were asked to speak short command and control utterances such as digits and spelling loops or operated applications for hands-free telephony, radio and navigation. The order of utterances and speakers was also randomized. The speaker change rate from one utterance to the next was set to about 7.5 %. At least 5 utterances were spoken between two speaker turns. The speaker specific codebooks were evaluated in parallel according to (5.23). The codebook of the target speaker was continuously adapted by combining EV and ML estimates as described by (4.10). Hence, speaker adaptation was supervised in the sense that the speaker identity was given to prevent any maladaptation.

All speakers of this subset are subsequently considered to describe the likelihood evolution only depending on the number of utterances used for adaptation N_{A}. Speaker dependencies of this evolution are not modeled for reasons of flexibility. A statistical description is targeted which does not depend on a specific speaker set. Subsequently, log-likelihood ratios with respect to the standard codebook are considered for target and non-target speakers. The standard codebook as the origin of all speaker specific codebooks is viewed as a reasonable reference.

---

[1] The USKCP is a speech database internally collected by TEMIC Speech Dialog Systems, Ulm, Germany. The USKCP comprises command and control utterances for in-car applications such as navigation commands, spelling and digit loops. The language is US-English.

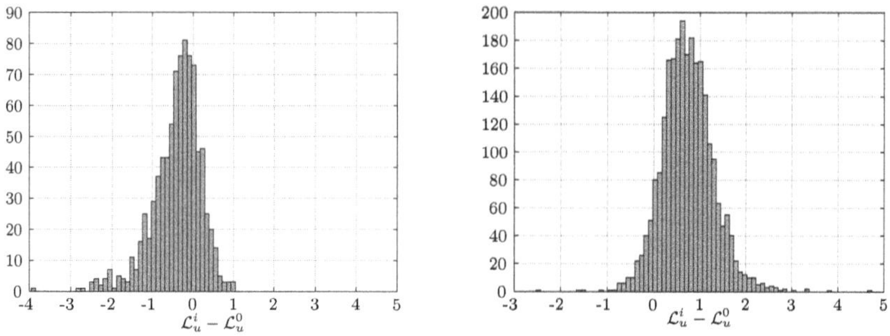

(a) Non-target speaker. (b) Target speaker.

Fig. 6.1 Histogram of the log-likelihood ratios $\mathcal{L}_u^i - \mathcal{L}_u^0$ for non-target speakers and the target speaker. $\lambda = 4$ and $N_A = 5$ are employed.

The central limit theorem predicts a Gaussian distribution of a random variable if realized by a sum of an infinite number of independent random variables [Bronstein et al., 2000; Zeidler, 2004]. Subsequently, it is assumed that this condition is approximately fulfilled since $\mathcal{L}_u^i$ is an averaged log-likelihood.

In Fig. 6.1 and Fig. 6.2 both the histogram and the approximation by univariate Gaussian distributions are shown for some examples. Asymmetric distributions, e.g. as found by Bennett [2003], are not considered for reasons of simplicity. In consequence univariate symmetric Gaussian distributions are used to represent the log-likelihood ratios. The following two cases are considered:

- **Target speaker.** The codebook under investigation belongs to the target speaker. This match is described by the parameter set $\{\dot{\mu}, \dot{\sigma}\}$ comprising mean and standard deviation.
- **Non-target speakers.** The codebook belongs to a non-target speaker profile. This mismatch situation is described by the parameter set $\{\ddot{\mu}, \ddot{\sigma}\}$.

To estimate the parameter set $\{\dot{\mu}, \dot{\sigma}\}$ several adaptation stages comprising 5 utterances are defined. All log-likelihoods $\mathcal{L}_u^i$ measured within a certain adaptation stage N_A are employed to estimate the mean and standard deviation of the pooled log-likelihood ratio scores. The parameters $\{\ddot{\mu}, \ddot{\sigma}\}$ of the non-target speaker models can be calculated in a similar way. The procedure only differs in the fact that speaker model and current speaker do not match [Herbig et al., 2010e].

In Figs. 6.3 - 6.5 the parameters $\{\dot{\mu}, \ddot{\mu}, \dot{\sigma}, \ddot{\sigma}\}$ are shown for certain adaptation stages N_A. Different values of the parameter λ were employed in speaker adaptation.

In summary, the Figs. 6.3 - 6.5 reveal a remarkable behavior of the speaker specific codebooks which should be discussed in more detail. The threshold λ for combining EV and ML estimates in (4.11) is increased from Fig. 6.3 to

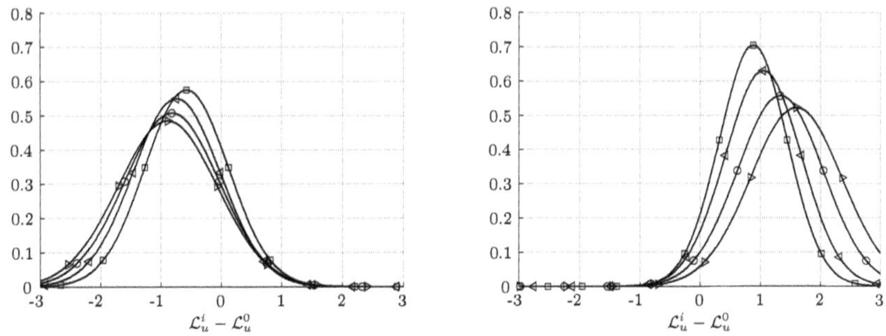

(a) Non-target speaker. (b) Target speaker.

Fig. 6.2 Distribution of log-likelihood ratios for dedicated adaptation stages - $N_\mathrm{A} = 10$ (□), $N_\mathrm{A} = 20$ (◁), $N_\mathrm{A} = 50$ (○) and $N_\mathrm{A} = 100$ (▷). $\lambda = 4$ is applied for codebook adaptation. Figure is taken from [Herbig et al., 2010e].

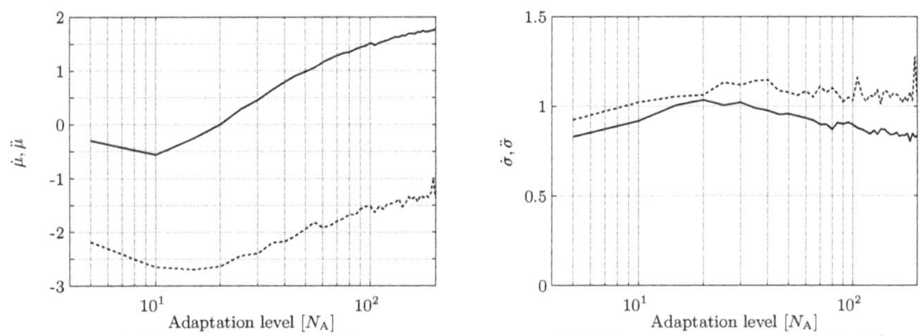

(a) Mean values $\{\dot{\mu}, \ddot{\mu}\}$. (b) Standard deviations $\{\dot{\sigma}, \ddot{\sigma}\}$.

Fig. 6.3 Log-likelihood ratios of the target (solid line) and non-target speaker models (dashed line) for different training levels. The convex combination in (4.10) and (4.11) uses only ML estimates ($\lambda \to 0$).

Fig. 6.5. In Fig. 6.3 this constant is almost zero. Only ML estimates are applied whereas in Fig. 6.5 pure EV estimates are employed.

At first glance the log-likelihood values behave as expected. The correct assignment of an utterance to the corresponding codebook yields a clear improvement with respect to the standard codebook. Log-likelihood ratios increase with continued speaker adaptation. Even with only a few utterances an improvement can be achieved on average. The prior knowledge incorporated into the EV estimates enables fast speaker adaptation based on only a few adaptation parameters.

However, one would not employ the scenario depicted in Fig. 6.3. During the first 20 utterances, the performance of the modified codebooks is significantly worse compared to the standard codebook. The initial mean vectors in (4.10) are substituted by unreliable ML estimates which converge slowly.

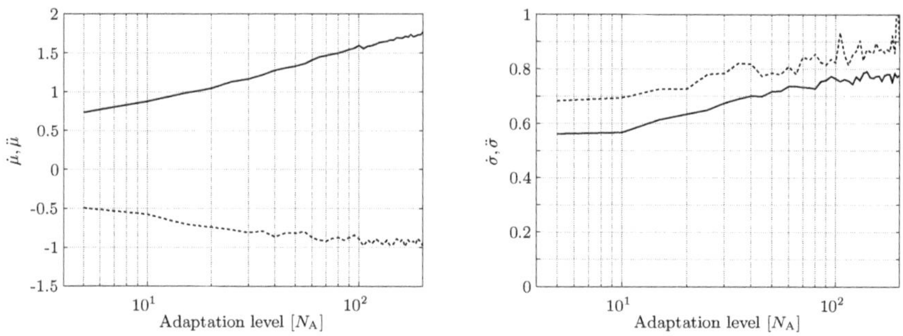

(a) Mean values $\{\dot{\mu}, \ddot{\mu}\}$. (b) Standard deviations $\{\dot{\sigma}, \ddot{\sigma}\}$.

Fig. 6.4 Log-likelihood ratios of the target (solid line) and non-target speaker models (dashed line) for different training levels. The convex combination in (4.10) and (4.11) uses $\lambda = 4$. Figure is taken from [Herbig et al., 2010c,e].

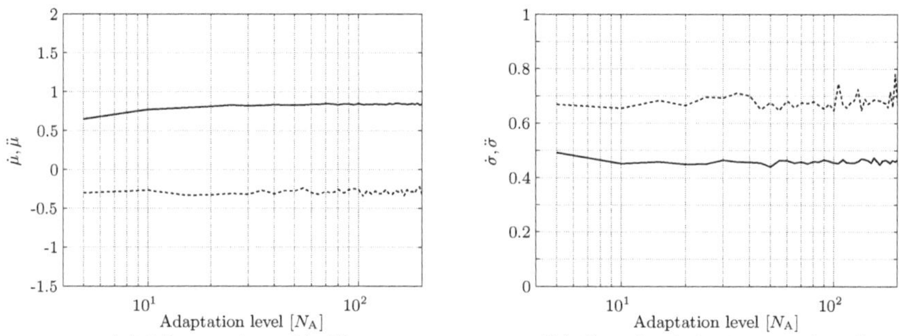

(a) Mean values $\{\dot{\mu}, \ddot{\mu}\}$. (b) Standard deviations $\{\dot{\sigma}, \ddot{\sigma}\}$.

Fig. 6.5 Log-likelihood ratios of the target (solid line) and non-target speaker models (dashed line) for different training levels. The convex combination in (4.10) and (4.11) only uses EV estimates ($\lambda \to \infty$).

In Fig. 6.5 the other extreme case is depicted. ML estimates are consequently neglected and only EV adaptation is applied. Fast adaptation is achieved on the first utterances but the log-likelihood scores settle after some $20 - 30$ utterances. This shows the restriction of this implementation of the EV approach since only 10 parameters can be estimated. This finding agrees with the reasoning of Botterweck [2001].

On average codebooks of non-target speakers perform worse than the standard codebook. When their codebooks are continuously adapted, the discrimination capability of speaker specific codebooks becomes obvious. The mismatch between statistical model and actual speaker characteristics is increasingly dominant. However, the extent of this drift is not comparable to the likelihood evolution observed for the target speaker.

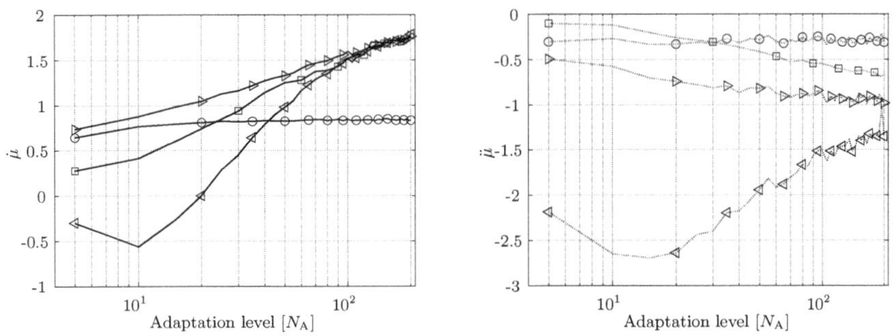

(a) Target speaker. (b) Non-target speaker.

Fig. 6.6 Evolution of log-likelihood ratios for different speaker adaptation schemes - combined EV and ML adaptation with $\lambda = 4$ ($\triangleright$), MAP adaptation with $\eta = 4$ ($\square$), EV ($\circ$) and ML ($\triangleleft$) adaptation.

Dedicated values of the means and standard deviations shown in Fig. 6.4 are represented in Fig. 6.2 by the corresponding univariate Gaussian distributions. In Fig. 6.6 likelihood evolution is evaluated for several adaptation schemes.

At a second glance these distributions reveal some anomalies and conflicts. The conclusion that the codebook of the target speaker performs better than the standard codebook and that codebooks of non-target speakers are worse is only valid on average. The investigation reveals a large overlap of both distributions. Especially a weakly adapted codebook of the target speaker seems to perform often either comparably well or even worse than non-target speaker models. However, even relatively well adapted models of non-target speakers might behave comparably or better than the standard codebook in some cases.

Obviously, it is complicated to construct a self-organizing system as long as the start or enrollment of new speakers is critical. As soon as about 20 utterances are assigned to the correct speaker model, this situation is less critical since the overlap is reduced. Especially in the learning phase, the use of posterior probabilities should help to prevent false decisions and error propagation. Those errors are usually not detected by speaker identification techniques when only ML estimates are employed.

The Figs. 6.3 - 6.5 contain a further detail concerning the standard deviations of the log-likelihood ratios. Larger standard deviations can be observed for continuously adapted codebooks. The combination of short-term and long-term speaker adaptation discussed in Sect. 4.2 might give a reasonable explanation. Speaker adaptation starts with a small set of adaptation parameters and allows a higher degree of individualism as soon as a sufficient amount of training data is available. This restriction might explain the relatively tight Gaussian distributions at the beginning. When N_A exceeds a threshold of about 10 utterances, the standard deviation increases significantly. However, the shift of the Gaussian distributions is larger than the broadening of the curve. A tendency towards more reliable decisions can be observed.

Finally, this approach should be discussed with respect to the literature. Yin et al. [2008] investigate the influence of speaker adaptation on speaker specific GMMs. They conclude that likelihood scores drift with an increasing length of enrollment duration. They assume Gaussian distributed likelihoods and perform mean subtraction and variance normalization to receive normalized scores. In contrast to this book, a decision strategy based on likelihoods instead of posterior probabilities is used.

In the following the finding of the experiment described in this section is used to calculate posterior probabilities which reflect the observed likelihood scores based on a single utterance and the training level of all speaker specific codebooks. In a second step this concept is extended to posterior probabilities which take a series of successive utterances into consideration.

6.2.2 Posterior Probability Computation at Run-Time

Prior knowledge about likelihood evolution allows calculating meaningful posterior probabilities which are more reliable than pure likelihoods as will be shown later.

In the following, posterior probabilities are derived for each speaker. They equally consider the observed likelihood scores given the current utterance $\mathsf{X}_u = \{\mathbf{x}_1, \ldots, \mathbf{x}_{T_u}\}$ and the training levels of all speaker models. This includes not only the log-likelihood $\mathcal{L}_u^{i_u}$ of the hypothesized target speaker i_u but also log-likelihoods of the standard codebook $\mathcal{L}_u^0$ and non-target speaker models $\mathcal{L}_u^{i \neq i_u}$. The performance of each speaker model is evaluated in the context of the codebooks of all enrolled speakers. The procedure has to be applicable to a varying number of enrolled speakers without any re-training since a self-learning speech controlled system is targeted [Herbig et al., 2010e].

The proposed posterior probability is decomposed into the contributions of each speaker model and is derived by applying Bayes' theorem. The posterior probability

$$p(i_u|\mathcal{L}_u^0, \mathcal{L}_u^1, \ldots, \mathcal{L}_u^{N_{\mathrm{Sp}}}) = \frac{p(\mathcal{L}_u^0, \mathcal{L}_u^1, \ldots, \mathcal{L}_u^{N_{\mathrm{Sp}}}|i_u) \cdot p(i_u)}{p(\mathcal{L}_u^0, \mathcal{L}_u^1, \ldots, \mathcal{L}_u^{N_{\mathrm{Sp}}})} \tag{6.1}$$

is represented by the conditional probability density of all log-likelihood scores, prior probability of the assumed speaker and a normalization.

However, it seems to be advantageous to use relative measures instead of the absolute likelihood scores:

- Absolute likelihood scores of speaker specific codebooks obviously do not provide a robust measure. For example, adverse environments might affect modeling accuracy. This effect is unwanted for speaker identification and can be circumvented or at least alleviated by using relative instead of absolute measures. Adverse conditions are expected to affect all codebooks in a similar way. The standard codebook as a speaker independent and extensively trained speech model can be considered as a reasonable reference. This problem is related to the problem of open-set speaker identification as found by Fortuna et al. [2005].

- Another strong point to employ relative scores is the likelihood evolution described in the last section. The likelihood converges to higher values when the statistical models are continuously adapted. The range of log-likelihood scores depends on the effective number of speaker adaptation parameters.
- Furthermore, the log-likelihoods of the statistical model representing the target speaker seem to be only marginally correlated with those of non-target speaker models if relative measures, e.g. $\mathcal{L}_u^i - \mathcal{L}_u^0$, are considered instead of $\mathcal{L}_u^i$. A similar result can be observed when speaker models of two non-target speakers are compared. In both cases the correlation coefficient [Bronstein et al., 2000] given by

$$\rho(a,b) = \frac{\mathrm{E_{a,b}}\{(a - \mathrm{E_a}\{a\}) \cdot (b - \mathrm{E_b}\{b\})\}}{\sqrt{\mathrm{E_a}\{(a - \mathrm{E_a}\{a\})^2\}} \cdot \sqrt{\mathrm{E_b}\{(b - \mathrm{E_b}\{b\})^2\}}} \tag{6.2}$$

$$a = \mathcal{L}^i - \mathcal{L}^0, \qquad i = 1, \ldots, N_{\mathrm{Sp}} \tag{6.3}$$

$$b = \mathcal{L}^j - \mathcal{L}^0, \qquad j = 1, \ldots, N_{\mathrm{Sp}},\ j \neq i \tag{6.4}$$

was quite small on average when all speakers of the development set were investigated. In the case of speaker models trained on a few utterances the magnitude of the correlation coefficient was $|\rho| < 0.2$, in general. A further decrease could be observed for moderately and extensively trained speaker models.

Thus, equation (6.1) is rewritten in terms of relative measures with respect to the score of the standard codebook:

$$p(i_u|\mathcal{L}_u^0, \mathcal{L}_u^1, \ldots, \mathcal{L}_u^{N_{\mathrm{Sp}}}) = \frac{p(\mathcal{L}_u^1, \ldots, \mathcal{L}_u^{N_{\mathrm{Sp}}}|\mathcal{L}_u^0, i_u) \cdot p(\mathcal{L}_u^0|i_u) \cdot p(i_u)}{p(\mathcal{L}_u^1, \ldots, \mathcal{L}_u^{N_{\mathrm{Sp}}}|\mathcal{L}_u^0) \cdot p(\mathcal{L}_u^0)}. \tag{6.5}$$

Subsequently, it is assumed that speaker specific log-likelihoods $\mathcal{L}^i$ which are normalized by $\mathcal{L}^0$ can be treated separately or equivalently can be viewed as statistically independent:

$$p(i_u|\mathcal{L}_u^0, \mathcal{L}_u^1, \ldots, \mathcal{L}_u^{N_{\mathrm{Sp}}})$$

$$= \frac{p(\mathcal{L}_u^1|\mathcal{L}_u^0, i_u)}{p(\mathcal{L}_u^1|\mathcal{L}_u^0)} \cdot \ldots \cdot \frac{p(\mathcal{L}_u^{N_{\mathrm{Sp}}}|\mathcal{L}_u^0, i_u)}{p(\mathcal{L}_u^{N_{\mathrm{Sp}}}|\mathcal{L}_u^0)} \cdot \frac{p(\mathcal{L}_u^0|i_u) \cdot p(i_u)}{p(\mathcal{L}_u^0)} \tag{6.6}$$

$$= \frac{p(\mathcal{L}_u^1|\mathcal{L}_u^0, i_u)}{p(\mathcal{L}_u^1|\mathcal{L}_u^0)} \cdot \ldots \cdot \frac{p(\mathcal{L}_u^{N_{\mathrm{Sp}}}|\mathcal{L}_u^0, i_u)}{p(\mathcal{L}_u^{N_{\mathrm{Sp}}}|\mathcal{L}_u^0)} \cdot p(i_u|\mathcal{L}_u^0). \tag{6.7}$$

Furthermore, the log-likelihoods of the standard codebook $\mathcal{L}_u^0$ are assumed to be independent of the speaker identity i_u:

$$p(i_u|\mathcal{L}_u^0, \mathcal{L}_u^1, \ldots, \mathcal{L}_u^{N_{\mathrm{Sp}}}) = \frac{p(\mathcal{L}_u^1|\mathcal{L}_u^0, i_u)}{p(\mathcal{L}_u^1|\mathcal{L}_u^0)} \cdot \ldots \cdot \frac{p(\mathcal{L}_u^{N_{\mathrm{Sp}}}|\mathcal{L}_u^0, i_u)}{p(\mathcal{L}_u^{N_{\mathrm{Sp}}}|\mathcal{L}_u^0)} \cdot p(i_u). \tag{6.8}$$

The denominators contain normalization factors which guarantee that the sum over all posterior probabilities $p(i_u|\mathcal{L}_u^0, \mathcal{L}_u^1, \ldots, \mathcal{L}_u^{N_{\mathrm{Sp}}})$ is equal to unity. Therefore, the final result can be given by

$$p(i_u|\mathcal{L}_u^0, \mathcal{L}_u^1, \ldots, \mathcal{L}_u^{N_{\mathrm{Sp}}}) \propto p(\mathcal{L}_u^1|\mathcal{L}_u^0, i_u) \cdot \ldots \cdot p(\mathcal{L}_u^{N_{\mathrm{Sp}}}|\mathcal{L}_u^0, i_u) \cdot p(i_u). \tag{6.9}$$

To compute the posteriori probability $p(i_u|\mathcal{L}_u^0, \mathcal{L}_u^1, \ldots, \mathcal{L}_u^{N_{\mathrm{Sp}}})$, it is assumed that the speaker under investigation is the target speaker whereas all remaining speakers are non-target speakers. Thus, the density function $p(\mathcal{L}_u^{i_u}|\mathcal{L}_u^0, i_u)$ has to be calculated in the case of the target speaker and $p(\mathcal{L}_u^{i \neq i_u}|\mathcal{L}_u^0, i_u)$ in the case of a non-target speaker. Both cases are combined in (6.9). This step is repeated for all speakers. This procedure is not limited by the number of enrolled speakers N_{Sp} because an additional speaker only causes a further factor in (6.9). Thus, each posterior probability takes all log-likelihoods of a particular utterance into consideration.

As discussed so far, it is not obvious how this approach reflects different training levels. In the following, the knowledge about the adaptation status is directly incorporated into the parameters of the conditional density functions $\left\{p(\mathcal{L}_u^1|\mathcal{L}_u^0, i_u), \ldots, p(\mathcal{L}_u^{N_{\mathrm{Sp}}}|\mathcal{L}_u^0, i_u)\right\}$.

In general, conditional Gaussian distributions [Bishop, 2007] seem to offer a statistical framework suitable to integrate the log-likelihood of a particular speaker and the reference given by the standard codebook. Alternatively, univariate Gaussian distributions can be applied which are trained and evaluated on the log-likelihood ratio $\mathcal{L}_u^i - \mathcal{L}_u^0$ as done in the preceding section. For reasons of simplicity the latter approach is employed.

Two sets of parameters are required for each adaptation level as discussed before. The parameters for the target speaker are characterized by the mean $\dot{\mu}$ and standard deviation $\dot{\sigma}$ based on log-likelihood ratios. The parameters for non-target speakers are denoted by $\ddot{\mu}$ and $\ddot{\sigma}$. Now the conditional density functions $p(\mathcal{L}_u^i|\mathcal{L}_u^0, i_u)$ can be calculated for the target speaker

$$p(\mathcal{L}_u^{i_u}|\mathcal{L}_u^0, i_u) = \mathcal{N}\left\{\mathcal{L}_u^{i_u} - \mathcal{L}_u^0|\dot{\mu}_{i_u}, \dot{\sigma}_{i_u}\right\} \tag{6.10}$$

and non-target speakers

$$p(\mathcal{L}_u^{i \neq i_u}|\mathcal{L}_u^0, i_u) = \mathcal{N}\left\{\mathcal{L}_u^{i \neq i_u} - \mathcal{L}_u^0|\ddot{\mu}_i, \ddot{\sigma}_i\right\}. \tag{6.11}$$

For testing the parameter sets $\{\dot{\mu}, \dot{\sigma}\}$ and $\{\ddot{\mu}, \ddot{\sigma}\}$ are calculated for each speaker individually. The parameters depend only on the number of training utterances N_{A}. With respect to Fig. 6.4 this problem can be considered as a regression problem. For example, a Multilayer Perceptron[2] (MLP) may be trained to interpolate the graph for unseen N_{A}. For each scenario one MLP was trained prior to the experiments which will be discussed later. Each MLP may be realized by one node in the input layer, 4 nodes in the hidden layer and two output nodes. The latter represent the mean and standard deviation of the univariate Gaussian distribution. Another strategy might be to categorize mean values and standard deviations in groups of several adaptation stages and to use a look-up table.

In summary, posterior probabilities are employed to reflect the match between the target speaker model and an observation as well as the expected log-likelihood due to the adaptation stage of the statistical model. In addition, the performance of non-target speaker models is evaluated likewise. The computation of the posterior probabilities can be simplified when all log-likelihoods are normalized by the log-likelihood of the standard codebook. The resulting posterior probability can be factorized. Prior knowledge about likelihood evolution is directly incorporated into the parameters of the corresponding univariate Gaussian distributions.

In the following sections long-term speaker tracking is described. The goal is to calculate a posterior probability which is not only calculated on a single utterance. Instead a series of successive utterances is investigated to determine the probability that speaker i has spoken the utterance $\mathbf{X}_u$ given the preceding and successive utterances.

6.3 Closed-Set Speaker Tracking

As discussed so far, only a posterior probability of a single utterance has been considered. However, the goal of this chapter is to provide the posterior probability $p(i_u|\mathsf{X}_{1:N_u})$ for each speaker i which reflects an entire series of utterances $\mathsf{X}_{1:N_u}$. Therefore a statistical model has to be found which allows long-term speaker tracking. Referring to Sect. 2.4.1 both discriminative and generative statistical models can be applied to distinguish between different speakers:

- Discriminative models such as an MLP can be trained to separate enrolled speakers and to detect speaker changes. However, the number of speakers is often unknown and speech data for an enrollment is usually not available for a self-learning speaker identification system. Furthermore, an enrollment is undesirable with respect to convenience.
- Generative statistical models are preferred in this context since enrolled speakers are individually represented. A solution is targeted which is independent from the composition of the speaker pool.

[2] A detailed description of MLPs, their training and evaluation can be found by Bishop [1996].

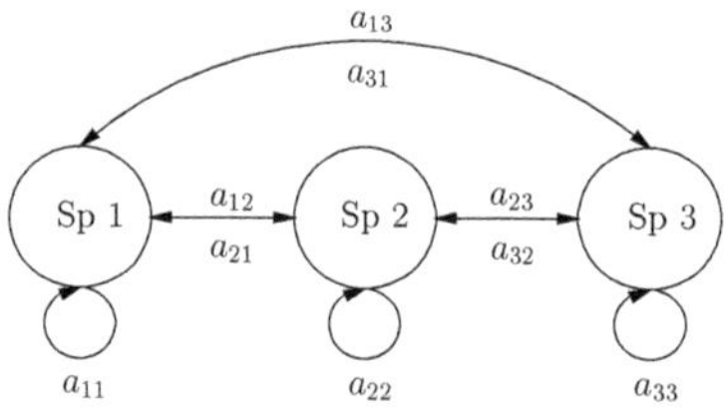

Fig. 6.7 Speaker tracking for three enrolled speakers realized by a Markov model of first order. The transitions a_{ij} denote the speaker change probability from speaker i to speaker j.

Subsequently, speaker tracking is viewed as a Markov process. Each state represents one enrolled speaker and the transitions model speaker changes from one utterance to the next. In Fig. 6.7 an example of three enrolled speakers is depicted. A flexible structure is used which allows the number of states to be increased with each new speaker until an upper limit is reached. Then one speaker model is dropped and replaced by a new speaker model to limit the computational complexity. Transition probabilities are chosen independently from enrolled speakers. They are either determined by an expected prior speaker change rate p_{Ch} or can be controlled by additional devices such as a beamformer if a discrete probability for speaker changes can be given. In contrast to Sect. 2.5.2, the Markov model is operated on asynchronous events namely utterances instead of equally spaced time instances.

In the next step the link between speaker identification described in Sect. 5.2 and the Markov model for speaker tracking has to be implemented. Speaker identification makes use of speaker specific codebooks which are interpreted as GMMs with uniform weighting factors.

A new HMM[3] can be constructed on an utterance level by combining codebooks and the Markov model for speaker tracking [Herbig et al., 2010e]. During speech recognition codebooks represent speaker specific pronunciations to guarantee optimal speech decoding. Now these codebooks are used to determine the average match with the speaker's characteristics for speaker identification according to (5.23). Codebooks are considered here in the context of a series of utterances and speaker changes. The emission probability density of a particular speaker is given by

$$p(\mathsf{X}_u|i_u) = \exp\left(T_u \cdot \mathcal{L}_u^{i_u}\right). \tag{6.12}$$

Each speaker or equivalently each state is described by an individual GMM and therefore the resulting HMM equals an CDHMM as introduced in

[3] A similar model the so-called segment model is known from the speech recognition literature. Instead of single observations $\mathbf{x}_t$, segments $\mathbf{x}_{1:T}$ of variable length T are assigned to each state to overcome some limitations of HMMs [Delakis et al., 2008].

Sect. 2.5.2. Decoding techniques for HMMs can be used to assign a series of utterances

$$\mathsf{X}_{1:N_\mathrm{u}} = \{\mathsf{X}_1, \ldots, \mathsf{X}_u, \ldots, \mathsf{X}_{N_\mathrm{u}}\} \tag{6.13}$$

to the most probable sequence of speaker identities

$$i_{1:N_\mathrm{u}}^{\mathrm{MAP}} = \{i_1^{\mathrm{MAP}}, \ldots, i_u^{\mathrm{MAP}}, \ldots, i_{N_\mathrm{u}}^{\mathrm{MAP}}\}. \tag{6.14}$$

This task can be solved by an iterative algorithm such as the *forward* algorithm already explained in Sect. 2.5.2. The forward algorithm can be employed to compute the joint probability $p(\mathsf{X}_{1:u}, i_u)$ of each speaker i_u given the observed history of utterances $\mathsf{X}_{1:u}$. The prior probability of each speaker is modeled here by $p(i_u) = \frac{1}{N_{\mathrm{Sp}}}$. N_{Sp} represents the number of enrolled speakers.

Additionally, the *backward* algorithm can be employed. It requires to save the likelihoods $p(\mathsf{X}_u|i_u)$ of N_u utterances in a buffer and to re-compute the complete backward path after each new utterance. For all speaker models the likelihood $p(\mathsf{X}_{u+1:N_\mathrm{u}}|i_u)$ has to be calculated as described in Sect. 2.5.2.

In Fig. 6.8 an example with 5 speakers and two exemplary paths are shown. The dashed path displays the correct path and the dotted one denotes a non-target speaker. The graphical representation in form of a finite state machine as used in Fig. 6.7 only discloses all possible states and transitions as well as the corresponding probabilities. The time trajectory is concealed. Thus, trellis representation is subsequently used because it spans all possible paths or equivalently the trajectory of states as a surface. The dashed path is iteratively processed by the forward algorithm. In this example the first three utterances are expected to produce high posterior probabilities since no speaker turn is presumed. After the speaker change there is a temporary decrease of the resulting posterior probability because speaker turns are penalized by the prior probability p_{Ch}. The backward algorithm encounters the inverse problem since only the last three utterances are expected to result in high posterior probabilities. This example shows the problem of incomplete posterior probabilities.

Therefore, complete posteriors are used to capture the entire temporal knowledge about the complete series of utterances. This is achieved by the fusion of forward and backward algorithm which was introduced in Sect. 2.5.2. Here only the final result is given:

$$p(i_u|\mathsf{X}_{1:N_\mathrm{u}}) \propto p(\mathsf{X}_{1:u}, i_u) \cdot p(\mathsf{X}_{u+1:N_\mathrm{u}}|i_u), \quad 0 < u < N_\mathrm{u} \tag{6.15}$$

$$p(i_{N_\mathrm{u}}|\mathsf{X}_{1:N_\mathrm{u}}) \propto p(\mathsf{X}_{1:N_\mathrm{u}}, i_{N_\mathrm{u}}). \tag{6.16}$$

The MAP criterion can be applied to determine the most probable speaker identity

$$i_u^{\mathrm{MAP}} = \arg\max_{i_u} \{p(i_u|\mathsf{X}_{1:N_\mathrm{u}})\}. \tag{6.17}$$

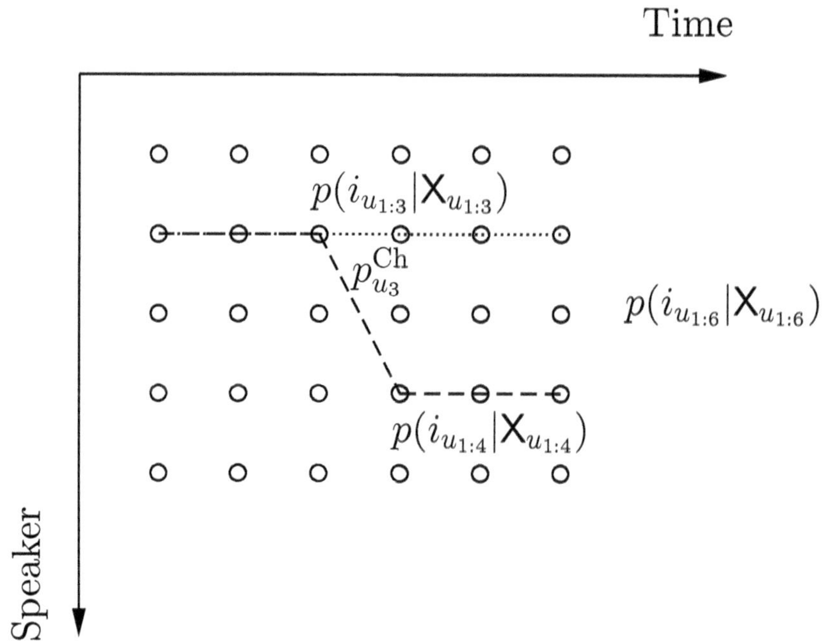

Fig. 6.8 Example for speaker tracking based on posterior probabilities. 5 speakers and two exemplary paths are shown. States are depicted versus the discrete time axis.

In addition, a direct measure for confidence or complementary uncertainty is achieved by posterior probabilities. If uncertainty exceeds a given threshold, particular utterances can be rejected by speaker identification. Those utterances are not used for speaker adaptation. In the experiments carried out adaptation was not performed when $p(i_u^{\mathrm{MAP}}|\mathsf{X}_{1:N_u}) < 0.5$.

The Markov model for speaker tracking and the modified posterior probability discussed before can be combined to obtain a more precise speaker modeling where likelihood evolution is considered. $p(\mathcal{L}_u^0, \mathcal{L}_u^1, \ldots, \mathcal{L}_u^{N_{\mathrm{Sp}}}|i_u)$ is employed in the forward-backward algorithm instead of the likelihood $p(\mathsf{X}_u|i_u)$. The former one can be obtained by applying Bayes' theorem to (6.9). The probability density function $p(\mathcal{L}_u^0, \mathcal{L}_u^1, \ldots, \mathcal{L}_u^{N_{\mathrm{Sp}}}|i_u)$ is used as an indirect measure for $p(\mathsf{X}_u|i_u)$ since the likelihoods $\left\{\mathcal{L}_u^0, \mathcal{L}_u^1, \ldots, \mathcal{L}_u^{N_{\mathrm{Sp}}}\right\}$ are derived from the observation X_u.

6.4 Open-Set Speaker Tracking

The focus of the preceding sections has been to enhance speech recognition by an unsupervised speaker tracking which is able to identify a limited number of enrolled speakers. Only the closed-set case has been investigated. The feature vectors of an utterance are only compared with existing speaker profiles and the one with the highest match or negatively spoken with lowest discrepancy is selected. Thus, the current user is always assigned to one of the enrolled speaker profiles. In this scenario speaker identification needs additional input when a new user operates the speech controlled device for the first time.

Long-term speaker tracking suffers from this deficiency in the same way. Although each speaker model is investigated with respect to its adaptation level, there is no measure whether the existing speaker models fit at all. The normalization in (6.16) always guarantees that the sum of all posterior probabilities equals unity. When there is a mismatch between an utterance and the speaker's codebook, this normalization can cause unwanted results and might pretend a high confidence. The goal is to define a statistical approach to detect mismatch conditions, especially unknown speakers.

In Sect. 2.4.3 some techniques were discussed to detect unknown speakers. The principle idea was to apply either fixed or dynamic thresholds to the log-likelihoods of each speaker model. Log-likelihood scores can be normalized by a reference, e.g. given by a UBM or speaker independent codebook, to obtain a higher robustness against false detections.

The Figs. 5.11, 6.3 - 6.5 and Fig. 6.9 show that a global threshold for log-likelihood ratios is difficult. The log-likelihood distributions significantly vary depending on the adaptation stage. Especially in the learning phase of a new codebook, it is complicated to find an optimal threshold. An unacceptably high error rate with respect to the discrimination of in-set and out-of-set speakers is expected.

Unknown speakers cannot be modeled directly because of a lack for training or adaptation data. However, an alternative model for situations when no speaker model represents an utterance appropriately can be obtained as a by-product of the posterior probability explained in Sect. 6.2.2.

In (6.9) each log-likelihood score is validated given the adaptation stage of the corresponding codebook. The range of log-likelihood values is known a priori. The main principle is to consider one speaker as the target speaker whereas remaining speakers have to be viewed as non-target speakers.

In the case of an unknown user no speaker model exists so that all speaker models act as non-target speaker models. In consequence, the posterior probability of an out-of-set speaker is proportional to the product of the univariate Gaussian functions of all non-target speakers [Herbig et al., 2011]:

$$p(i_u = N_{\mathrm{Sp}}+1|\mathcal{L}_u^0, \mathcal{L}_u^1, \ldots, \mathcal{L}_u^{N_{\mathrm{Sp}}}) \propto \prod_{i=1}^{N_{\mathrm{Sp}}} \mathcal{N}\left\{\mathcal{L}_u^i - \mathcal{L}_u^0 | \ddot{\mu}_i, \ddot{\sigma}_i\right\} \cdot p(i_u = N_{\mathrm{Sp}}+1). \tag{6.18}$$

The closed-set solution in (6.9) already contains the log-likelihood ratios of all possible non-target speaker models. Therefore, only the normalization has to be adjusted to guarantee that the sum over all in-set and out-of-set speakers equals unity.

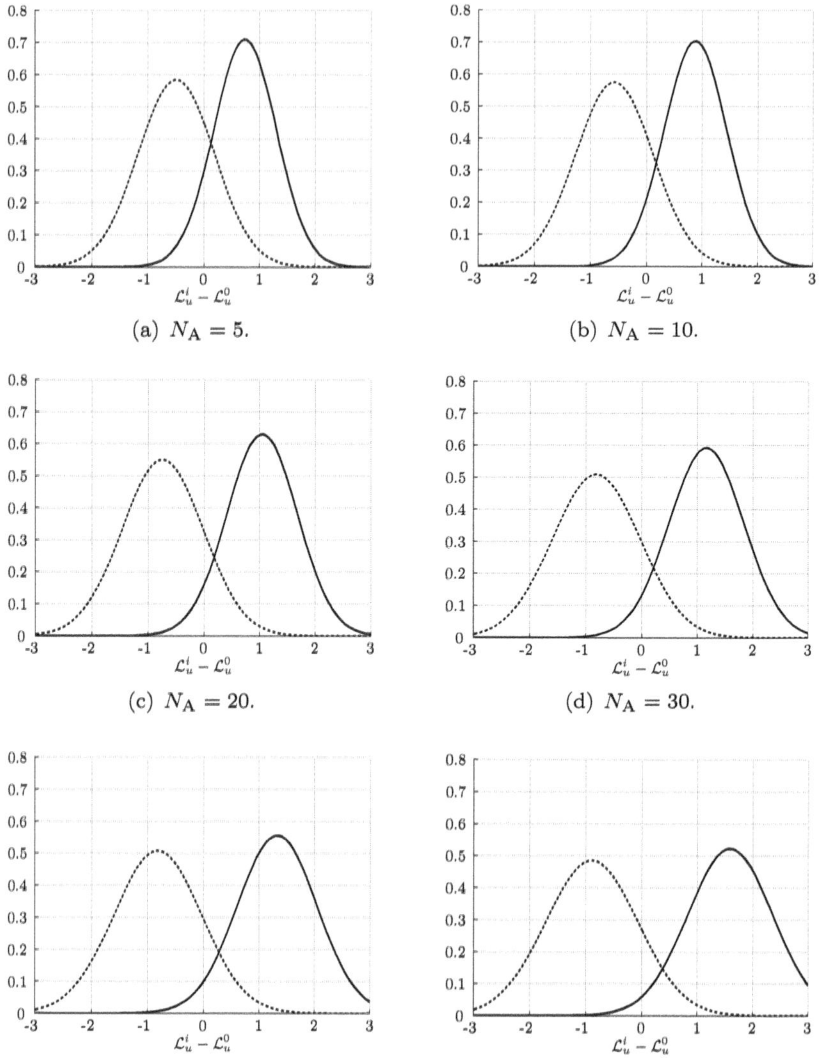

(a) $N_{\text{A}} = 5$.

(b) $N_{\text{A}} = 10$.

(c) $N_{\text{A}} = 20$.

(d) $N_{\text{A}} = 30$.

(e) $N_{\text{A}} = 50$.

(f) $N_{\text{A}} = 100$.

Fig. 6.9 Overlap of the log-likelihood distributions of the target (solid line) and non-target speaker models (dashed line). $\lambda = 4$ is employed in speaker adaptation. The x-coordinate denotes the log-likelihood ratio of speaker specific codebooks and standard codebook.

The concept of long-term speaker tracking, however, exploits the knowledge of a series of utterances and is more robust compared to a binary classifier based on likelihoods or posterior probabilities originating from a single utterance. The Markov model used for speaker tracking can be easily

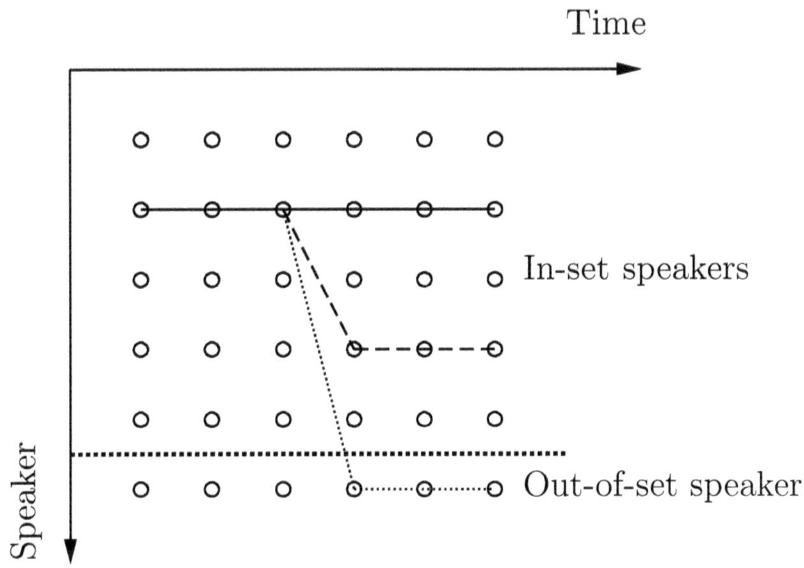

Fig. 6.10 Example for open-set speaker tracking.

extended from N_{Sp} to $N_{\mathrm{Sp}} + 1$ states to integrate the statistical modeling of unknown speakers into the forward-backward algorithm.

In Fig. 6.10 open-set speaker tracking is shown for an example of three paths. Two belong to enrolled speakers and an additional one represents an unknown speaker.

The MAP estimate in (6.17) may be used to detect unknown speakers in direct comparison with all enrolled speaker models. If the out-of-set speaker obtains the highest posterior probability, a new speaker specific codebook may be initialized. However, a relatively small posterior probability might be sufficient to enforce a new codebook. This can be the case when all posterior probabilities tend towards a uniform distribution.

It seems to be advantageous to consider speaker identification as a two-stage decision process [Angkititrakul and Hansen, 2007]. In the first step, the MAP criterion is applied as given in (6.17) to select the most probable in-set speaker. In the second step, this speaker identity has to be verified. This leads to a binary decision between in-set and out-of-set speakers. Hypotheses H_0 and H_1 characterize the event of an in-set and an out-of-set speaker, respectively. The optimal decision is defined in (2.14).

In general, the following two probabilities can be calculated by the forward-backward algorithm. The posterior probability

$$p(H_0|\mathsf{X}_{1:N_u}) = p(i_u \neq N_{\mathrm{Sp}} + 1|\mathsf{X}_{1:N_u}) \qquad (6.19)$$

denotes an enrolled speaker and the probability of an unknown is given by

$$p(H_1|\mathsf{X}_{1:N_u}) = p(i_u = N_{Sp} + 1|\mathsf{X}_{1:N_u}). \tag{6.20}$$

Optimal Bayes decisions depend on the choice of the costs for both error scenarios as described in Sect. 2.3. Subsequently, equal costs for not detected out-of-set speakers and not detected in-set speakers are assumed to simplify the following considerations. Bayes' theorem permits the following notation

$$p(\mathsf{X}_{1:N_u}|H_0) \cdot p(H_0) = p(H_0|\mathsf{X}_{1:N_u}) \cdot p(\mathsf{X}_{1:N_u}) \tag{6.21}$$

$$p(\mathsf{X}_{1:N_u}|H_1) \cdot p(H_1) = p(H_1|\mathsf{X}_{1:N_u}) \cdot p(\mathsf{X}_{1:N_u}), \tag{6.22}$$

where the posterior probabilities $p(H_0|\mathsf{X}_{1:N_u})$ and $p(H_1|\mathsf{X}_{1:N_u})$ are complementary

$$p(H_0|\mathsf{X}_{1:N_u}) + p(H_1|\mathsf{X}_{1:N_u}) = 1. \tag{6.23}$$

According to equation (2.14) the following criterion can be applied to decide whether a new speaker profile has to be initialized:

$$\frac{p(H_0|\mathsf{X}_{1:N_u})}{p(H_1|\mathsf{X}_{1:N_u})} \overset{H_1}{\leq} 1 \tag{6.24}$$

$$\frac{1 - p(H_1|\mathsf{X}_{1:N_u})}{p(H_1|\mathsf{X}_{1:N_u})} \overset{H_1}{\leq} 1 \tag{6.25}$$

$$p(H_1|\mathsf{X}_{1:N_u}) \overset{H_1}{\geq} \frac{1}{2}. \tag{6.26}$$

In particular, the posterior probabilities derived in Sect. 6.2.2 can be used in the forward-backward algorithm. Discrete decisions on an utterance level are avoided and a maximum of information is exploited. Furthermore, a series of utterances is considered to reduce the risk of false decisions. This technique provides a solution for the problem of open-set speaker tracking, especially when speaker models are differently trained.

6.5 System Architecture

In combination with long-term speaker tracking can a completely unsupervised system be implemented where the detection of unknown speakers is integrated [Herbig et al., 2011]. Since speech recognition requires a guess of the speaker identity on different time scales, the structure of the speaker identification process shown in Fig. 6.11 is employed.

I **Speech recognition.** Codebook selection is realized by a local speaker identification on a frame level as discussed in Sect. 5.2.1.

II **Preliminary speaker identification.** After each utterance the updates of the speaker specific parameter sets for energy normalization and mean subtraction are kept if no speaker turn is detected. Otherwise a reset is

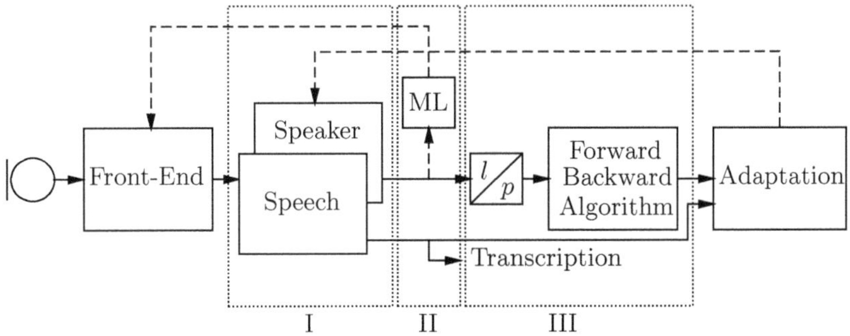

Fig. 6.11 System architecture for joint speaker identification and speech recognition comprising three stages. Part I and II denote the speaker specific speech recognition and preliminary speaker identification, respectively. In Part III posterior probabilities are calculated for each utterance and long-term speaker tracking is performed. Speaker adaptation is employed to enhance speaker identification and speech recognition. Codebooks are initialized in the case of an unknown speaker and the statistical modeling of speaker characteristics is continuously improved.

performed and the parameter set of the identified speaker is used subsequently. This stage is required for practical reasons since speaker identification and speech recognition accuracy are sensitive to an appropriate feature normalization.

III **Long-term speaker tracking.** At the latest after a predetermined number of utterances a final speaker identification is enforced. Speaker adaptation is then calculated for the most probable alignment of speaker identities. Speaker tracking is based on posterior probabilities originating from the reinterpretation of the observed log-likelihood scores. Unknown speakers can be detected. An additional speaker change detection, e.g. with the help of a beamformer, may be included without any structural modifications.

The speaker adaptation scheme presented in Chapter 4 allows the system to delay the final speaker identification since only the sufficient statistics of the standard codebook given by (4.14) and (4.15) are required. The assignment of observed feature vectors to the Gaussian distributions of a codebook is determined by the standard codebook. Therefore the speaker identity is not needed to accumulate the adaptation data. As soon as the speaker identity is determined, the data of this speaker is accumulated and an enhanced codebook is computed. To balance speaker identification accuracy with the latency of speaker adaptation and memory consumption, at most 6 successive utterances are subsequently employed for speaker tracking.

6.6 Evaluation

In the following experiments speaker identification has been extended by long-term speaker tracking. A completely unsupervised speech controlled system has been evaluated.

First, the closed-set scenario examined in Sect. 5.4.1 is discussed. Speaker identification and speech recognition are evaluated for speaker groups of 5 or 10 enrolled speakers [Herbig et al., 2010d,e]. In addition, the robustness of both implementations against speaker changes is examined.

Then speaker tracking is tested in a completely unsupervised way to simulate an open-set test case [Herbig et al., 2011]. The speech recognition rate is compared to the results of the first solution presented in Chapter 5 and optimal speaker adaptation investigated in Chapter 4.

6.6.1 *Closed-Set Speaker Identification*

In a first step, closed-set experiments are considered to investigate the speaker identification and speech recognition rates for speaker sets with 5 or 10 speakers [Herbig et al., 2010d,e]. Finally, the effect of speaker changes on the identification accuracy is examined.

Unsupervised Speaker Identification of 5 *enrolled Speakers*

In Sect. 5.4.1 closed-set identification was examined for groups of 5 speakers. In total, 60 different sets were evaluated. 75,000 test utterances were examined. The optimum of the WA was determined in Sect. 4.3 by supervised speaker adaptation in the sense that the speaker identity was given. This experiment has been repeated with long-term speaker tracking under realistic conditions to obtain a direct comparison.

The results are summarized in Table 6.1. The identification accuracy is noticeably improved by long-term speaker tracking compared to the results in Table 5.1. For $\lambda = 4$ a relative error rate reduction of approximately 74 % can be achieved by long-term speaker tracking. Unfortunately, speech recognition does not benefit from this approach and remains at the same level as before.

Again, the combination of EV and ML estimates yields the best identification result for $\lambda = 4$. 98.59 % of the speakers are correctly identified. For higher values of λ the identification accuracy is degraded. However, for all implementations except the baseline and short-term adaptation the identification rates are higher than for the experiments discussed in Sect. 5.4.1.

Even for $\lambda \approx 0$ an improvement can be stated since the temporary decrease of the likelihood shown in Fig. 6.3 is taken into consideration. For $\lambda \to \infty$ only a marginal improvement is achieved which can be explained by Fig. 6.5. Since the corresponding likelihood scores settle after $10 - 20$ utterances, no improvement can be expected by the knowledge of the adaptation status in the long run.

Table 6.1 Comparison of different adaptation techniques for self-learning speaker identification **with long-term speaker tracking**. Speaker sets comprising **5 enrolled speakers** are considered. Table is taken from [Herbig et al., 2010e].

Speaker adaptation	Rejected [%]	WA [%]	Speaker ID [%]
Baseline	-	85.23	-
Short-term adaptation	-	86.13	-
EV-MAP adaptation			
ML ($\lambda \approx 0$)	2.09	86.85	86.74
$\lambda = 4$	2.12	88.09	**98.59**
$\lambda = 8$	2.11	88.17	97.91
$\lambda = 12$	2.11	**88.21**	97.37
$\lambda = 16$	2.13	88.18	97.28
$\lambda = 20$	2.12	88.20	95.04
EV ($\lambda \to \infty$)	2.28	87.48	86.58
Typical errors			
min		± 0.23	± 0.09
max		± 0.25	± 0.25

Table 6.2 Comparison of different adaptation techniques for **long-term speaker tracking**. Speaker sets with **5 enrolled speakers** are investigated. Speaker identification and speech recognition results are presented for several evaluation bands - $\mathrm{I} = [1; 50[$, $\mathrm{II} = [50; 100[$, $\mathrm{III} = [100; 250]$ utterances.

Speaker adaptation	I		II		III	
	WA [%]	ID [%]	WA [%]	ID [%]	WA [%]	ID [%]
Baseline	84.51	-	86.84	-	85.02	-
Short-term adaptation	85.89	-	87.22	-	85.87	-
EV-MAP adaptation						
ML ($\lambda \approx 0$)	85.38	81.95	87.49	87.05	87.30	88.22
$\lambda = 4$	87.40	**97.82**	89.17	**98.29**	88.05	**98.94**
$\lambda = 8$	87.59	97.04	89.29	97.48	88.06	98.35
$\lambda = 12$	87.50	96.68	**89.39**	96.67	88.13	97.83
$\lambda = 16$	**87.62**	96.35	89.30	96.69	88.07	97.79
$\lambda = 20$	87.53	93.92	89.21	94.51	**88.17**	95.60
EV ($\lambda \to \infty$)	87.32	86.84	88.58	86.69	87.19	86.45
Typical errors						
min	± 0.53	± 0.24	± 0.49	± 0.21	± 0.30	± 0.10
max	± 0.58	± 0.63	± 0.54	± 0.54	± 0.33	± 0.32

Comparing the evaluation bands in Table 5.2 and Table 6.2 no significant differences with respect to speech recognition can be observed. Speaker identification, however, benefits from long-term speaker tracking even in the

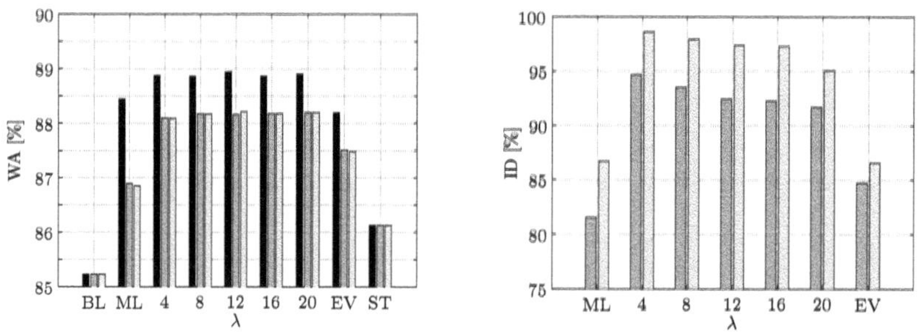

(a) Results for speech recognition. (b) Results for speaker identification.

Fig. 6.12 Comparison of supervised speaker adaptation with predefined speaker identity (black), joint speaker identification and speech recognition (dark gray) and long-term speaker tracking (light gray). Furthermore, the speaker independent baseline (BL) and short-term adaptation (ST) are shown. Figure is taken from [Herbig et al., 2010e].

learning phase given by the evaluation band I. This indicates that error propagation in the first evaluation band negatively affected the results of the implementations examined in the preceding chapter. Since this example has been discussed for several implementations, the results for supervised speaker adaptation, joint speaker identification and speech recognition and the combination with long-term speaker tracking are finally compared in Fig. 6.12. Both the WA and identification rate are depicted.

Unsupervised Speaker Identification of 10 *enrolled Speakers*

As discussed so far, speaker sets of 5 enrolled speakers have been evaluated. Even though 5 enrolled speakers seem to be sufficient for many use cases, e.g. in-car applications, the robustness of the speech controlled system should be investigated when the number of speakers is doubled. A new test set comprising 30 speaker sets is used for evaluation.

First, the joint speaker identification and speech recognition is tested without long-term speaker tracking to determine the performance decrease for speaker identification and speech recognition. Then, long-term speaker tracking is activated to evaluate whether an increased error rate on single utterances severely affects speaker tracking. The results are given in Table 6.3 and Table 6.4, respectively.

When Table 5.1 and Table 6.3 are compared despite different test sets, it can be stated that the speaker identification rates drop significantly. However, even with 10 enrolled speaker about 90 % identification rate can be achieved. In both cases about 20 % relative error rate reduction can be obtained for speech recognition when $\lambda = 4$ is used for adaptation. Thus, the speech recognition accuracy of the unified approach appears to be relatively insensitive to moderate speaker identification errors as expected [Herbig et al., 2010d].

Table 6.3 Comparison of different adaptation techniques for self-learning speaker identification **without long-term speaker tracking**. Speaker sets with **10 enrolled speakers** are investigated. Table is taken from [Herbig et al., 2010d].

Speaker adaptation	Rejected [%]	WA [%]	Speaker ID [%]
Baseline	-	84.93	-
Short-term adaptation	-	85.74	-
EV-MAP adaptation			
ML ($\lambda \approx 0$)	1.74	86.48	72.47
$\lambda = 4$	1.65	**87.96**	**89.56**
$\lambda = 8$	1.64	87.87	87.02
$\lambda = 12$	1.70	87.82	85.72
$\lambda = 16$	1.64	87.86	84.82
$\lambda = 20$	1.68	87.78	83.53
EV ($\lambda \to \infty$)	4.46	87.04	72.68
Typical errors			
min		± 0.23	± 0.22
max		± 0.26	± 0.33

Table 6.4 Comparison of different adaptation techniques for self-learning speaker identification **with long-term speaker tracking**. Speaker sets comprising **10 enrolled speakers** are investigated. Table is taken from [Herbig et al., 2010d].

Speaker adaptation	Rejected [%]	WA [%]	Speaker ID [%]
Baseline	-	84.93	-
Short-term adaptation	-	85.74	-
EV-MAP adaptation			
ML ($\lambda \approx 0$)	1.75	86.67	85.55
$\lambda = 4$	1.65	87.89	**97.89**
$\lambda = 8$	1.71	87.91	97.35
$\lambda = 12$	1.72	**88.05**	96.46
$\lambda = 16$	1.70	88.01	96.32
$\lambda = 20$	1.70	87.84	91.89
EV ($\lambda \to \infty$)	1.81	87.04	77.08
Typical errors			
min		± 0.23	± 0.10
max		± 0.26	± 0.30

However, a comparable level of the speaker identification rates to former experiments is achieved when speaker tracking is active as shown in Table 6.4.

Especially for larger speaker sets, a fast and robust information retrieval already in the learning phase is essential. With respect to Table 6.3 and Table 6.4 it becomes evident that an increased number of maladaptations also includes a less confident codebook selection which reduces the WA. Therefore long-term speaker tracking generally yields an advantage for the speaker identification and speech recognition rate.

For the considered use case ($N_{\mathrm{Sp}} < 10$) it can be concluded that a speech controlled system has been realized which is relatively insensitive to the number of enrolled speakers or the tuning parameter of speaker adaptation. Remarkably high identification rates have been achieved.

Effects of Speaker Changes on the Speaker Identification Accuracy

The techniques for speech controlled systems examined here do not use an explicit speaker change detection such as BIC. Instead, speaker changes are tracked by speaker identification on different time scales since speaker turns within an utterance are not considered. Nevertheless the question remains how the speech recognizer with integrated speaker identification reacts to speaker changes.

Codebooks are identified on a frame level for speaker specific speech recognition. A rapid switch between the codebooks of the enrolled speakers is possible if a clear mismatch between expected speaker characteristics and observed data is detected. But feature extraction is also speaker specific and is controlled on an utterance level. A reliable guess of the current speaker identity is available as soon as the entire utterance is processed. Since only moderate speaker change rates are expected, a parallel feature extraction can be avoided. This technique bears the risk not to detect speaker changes accurately since the parameter set of the speaker previously detected is used.

Subsequently, the vulnerability of speaker identification with respect to speaker changes is investigated. The prior probability of speaker changes $p_{\mathrm{Ch}} = 0.1$ is rather small and is comparable to the error rate of speaker identification on an utterance level as given in Table 5.1. It is assumed that most of the identification errors occur after speaker changes.

This question should be answered by certain points on the ROC curve which are shown in Fig. 6.13 and Fig. 6.14. The true positives that means correctly detected speaker turns are plotted versus the rate of falsely assumed speaker changes. Additional information about ROC curves can be found in Sect. A.3.

Fig. 6.13 and Fig. 6.14 are based on the results of Table 5.1 and Table 6.1, respectively. The performance of the target system either with or without long-term speaker tracking is depicted. Speaker change detection is given by the identification result of successive utterances. According to Fig. 6.13 and Fig. 6.14 it seems to be obvious that speaker changes are not a critical issue for the self-learning speaker identification. In fact, the evolution of the unsupervised system tends towards an even more robust system. But Fig. 6.13 and Fig. 6.14 also reveal that long-term speaker tracking only reduces the false alarm rate and even marginally increases the rate of missed speaker changes. The latter case can be explained by the fact that speaker changes are penalized by p_{Ch} in the forward-backward algorithm. In addition, the first utterance after a speaker change uses improper parameters for feature extraction.

In addition, several experiments for speaker change detection as described in Sect. 2.3 were conducted. Furthermore, the Arithmetic-Harmonic Sphericity (AHS) measure as found by Bimbot et al. [1995]; Bimbot and Mathan

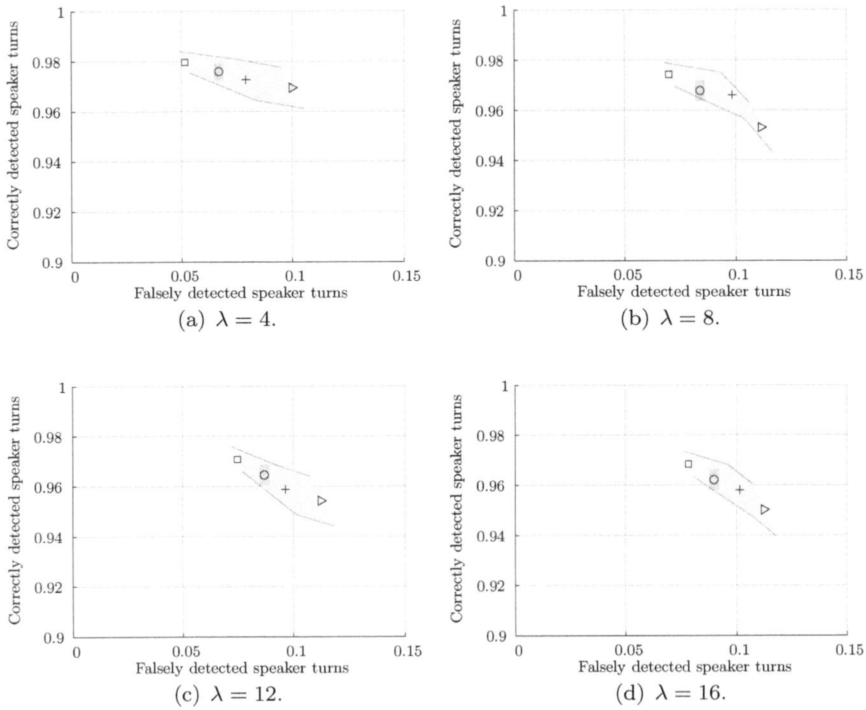

Fig. 6.13 Error rates of speaker change detection. The detection rate of speaker changes is plotted versus false alarm rate for *speaker identification* on an *utterance level*. The evaluation bands $\mathrm{I} = [1; 50[$, $\mathrm{II} = [50; 100[$, $\mathrm{III} = [100; 250]$ and the complete test set are used - I ($\triangleright$), II ($+$), III ($\Box$) and on average ($\circ$). Confidence bands are given by a gray shading.

[1993] was examined to capture speaker turns. Unfortunately, no robust speaker change detection could be achieved. The reason may be the interference of speaker and channel characteristics as well as background noises. It seems to be complicated to accurately detect speaker changes in adverse environments using simple statistical models as already mentioned in Sect. 2.3.2. Hence, only a prior probability for speaker changes was used for long-term speaker tracking. Speaker change detection was delegated to speaker identification.

6.6.2 Open-Set Speaker Identification

The closed-set scenario is now extended to an open-set scenario. As discussed so far, the first two utterances of a new speaker were indicated to belong to an unknown speaker as depicted in Fig.4.1(a).

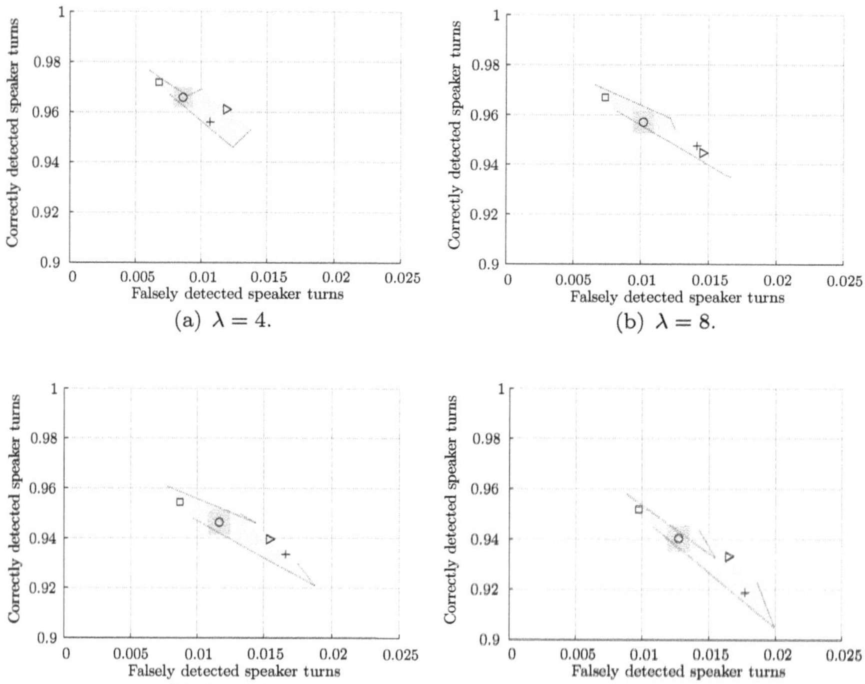

(a) $\lambda = 4$. (b) $\lambda = 8$. (c) $\lambda = 12$. (d) $\lambda = 16$.

Fig. 6.14 Error rates of speaker change detection. The detection rate of speaker changes is plotted versus false alarm rate for *long-term speaker tracking*. The evaluation bands I = $[1; 50[$, II = $[50; 100[$, III = $[100; 250]$ and the complete test set are used - I ($\triangleright$), II ($+$), III ($\square$) and on average ($\circ$). Confidence bands are given by a gray shading.

To obtain a strictly unsupervised speech controlled system, no information of the first occurrences of a new speaker is given by external means. If a known user is detected, the correct speaker profile has to be adjusted. Otherwise a new speaker profile has to be initialized on the first few utterances and continuously adapted on the subsequent utterances. The test setup is shown in Fig. 4.1(b).

Subsequently, long-term speaker tracking is extended so that unknown speakers can be detected as described in Sect. 6.4. The evaluation set comprises 5 speakers in each speaker set and is identical to the experiment repeatedly discussed before. To limit memory consumption and computational load, only one out-of-set and at most six in-set speakers can be tracked. If necessary one profile, e.g. the most recently initialized speaker model, is replaced by a new one.

Table 6.5 contains the results measured on the complete test set. A more detailed investigation of several evaluation bands is given in Table 6.6. Again,

Table 6.5 Comparison of different adaptation techniques for an **open-set** scenario. Speaker sets with **5 speakers** are examined. Table is taken from [Herbig et al., 2011].

Speaker adaptation	WA [%]
Baseline	85.23
Short-term adaptation	86.13
EV-MAP adaptation	
ML ($\lambda \approx 0$)	86.73
$\lambda = 4$	87.64
$\lambda = 8$	87.74
$\lambda = 12$	87.73
$\lambda = 16$	87.70
$\lambda = 20$	**87.77**
EV ($\lambda \to \infty$)	87.20
Typical errors	
min	±0.23
max	±0.25

Table 6.6 Comparison of different adaptation techniques for an **open-set** scenario. Speaker sets with **5 speakers** are investigated. The speech recognition rate is examined on evaluation bands comprising I = [1; 50[, II = [50; 100[, III = [100; 250] utterances.

Speaker adaptation	I WA [%]	II WA [%]	III WA [%]
Baseline	84.51	86.84	85.02
Short-term adaptation	85.89	87.22	85.87
EV-MAP adaptation			
ML ($\lambda \approx 0$)	85.36	87.08	87.24
$\lambda = 4$	86.61	88.75	87.74
$\lambda = 8$	86.84	**88.86**	87.78
$\lambda = 12$	**86.99**	88.84	87.70
$\lambda = 16$	86.85	88.83	87.72
$\lambda = 20$	86.88	88.81	**87.83**
EV ($\lambda \to \infty$)	86.84	88.32	86.98
Typical errors			
min	±0.54	±0.50	±0.30
max	±0.58	±0.54	±0.33

the combination of EV and ML estimates outperforms pure ML or EV estimates as well as the baseline and short-term adaptation without speaker identification. The parameter λ can be selected in a relatively wide range without negatively affecting speech recognition.

In summary, the achieved WA is smaller compared to closed-set experiments but there is still a clear improvement with respect to the baseline and short-term adaptation. Due to unsupervised speaker identification the recognition rate was reduced from 88.87 % (Table 4.1) to 88.10 % WA (Table 5.1) for $\lambda = 4$. The additional decrease to 87.64 % WA therefore demonstrates how important the training of new speaker models based on the first two utterances and the knowledge about the number of users seem to be for the experiments described in the preceding sections.

The performance of the closed-set and open-set implementation are compared in Fig. 6.15. On the first evaluation band I a relative increase of 6.4 % can be observed for the error rate of the open-set implementation when $\lambda = 4$ is used for codebook adaptation. On the evaluation bands II and III, however, the open-set implementation approaches the closed-set realization. Here relative error rates of 4 % and 2.7 % can be measured. Comparing Table 5.2 with Table 6.6 it can be stated that this effect is not limited to $\lambda = 4$. This finding gives reason to assume long-term stability of the unsupervised system.

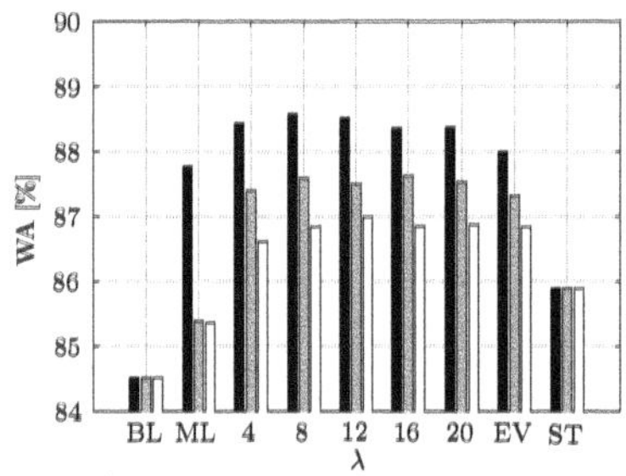

(a) Evaluation band I = [1; 50[utterances.

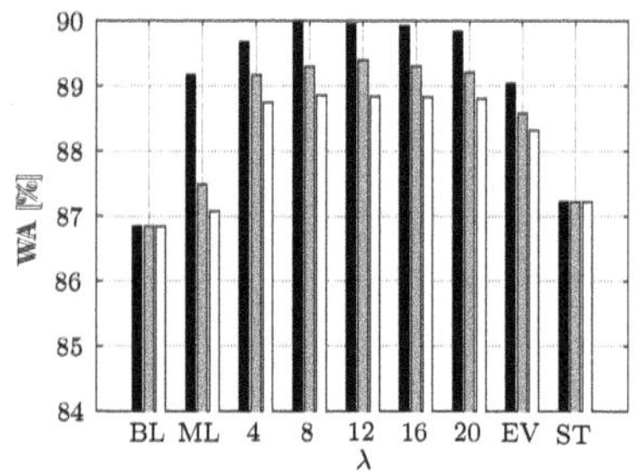

(b) Evaluation band II = [50; 100[utterances.

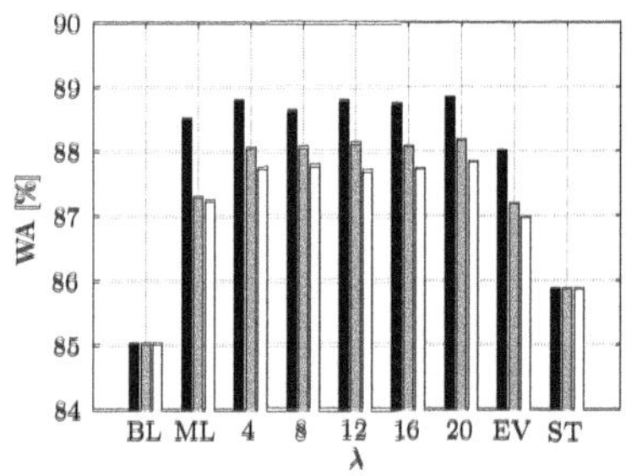

(c) Evaluation band III = [100; 250] utterances.

Fig. 6.15 Speech recognition is compared for supervised speaker adaptation with predefined speaker identity (black) and long-term speaker tracking for the closed-set (gray) and open-set (white) scenario. Improvements in speech recognition rate are evaluated with respect to the baseline (BL) and short-term adaptation (ST) since different subsets have been examined. Figure is taken from [Herbig et al., 2011].

6.7 Summary

The unified approach for speaker identification and speech recognition introduced in the preceding chapter can be extended by an additional component to appropriately track speakers on a series of utterances. Speaker adaptation can be delayed until a more confident guess of the speaker identity is obtained. The main aspect is to consider speaker profiles with respect to their training and to provide prior knowledge about their expected performance. A method has been developed to compute posterior probabilities reflecting not only the value of the log-likelihood but also prior knowledge about the evolution of speaker profiles. The performance of non-target speaker profiles is included to calculate a probability for each enrolled speaker to be responsible for the current utterance. By using a simple HMM on an utterance level, the decoding techniques known from speech recognition can be applied to determine the most probable speaker alignment.

A significantly higher speaker identification rate compared to the first realization of the target system was achieved for closed sets. An optimum of 98.59 % correctly identified speakers was obtained for $\lambda = 4$ whereas an identification rate of 94.64 % was observed without long-term tracking.

Long-term speaker tracking is essential to implement a speaker specific speech recognizer operated in an unsupervised way. Since this technique can be easily extended to open-set speaker tracking, new speaker profiles can be initialized without any additional intervention of new users. No training or explicit authentication of a new speaker is required.

Without long-term speaker tracking a notable number of codebooks is expected to be initialized even though an enrolled user is speaking as the corresponding ROC curves in Sect. 5.4.2 suggest. When memory is limited, only a small number of codebooks can be considered. Only weakly or moderately speaker profiles would be effectively used for speech recognition if unknown speakers were not reliably detected. On the other hand missed speaker changes would also negatively affect speech and speaker modeling leading to significantly higher recognition error rates. Applying long-term speaker tracking can help to prevent a decrease of the speech recognizer's performance. A steady stabilization of the complete system was observed in the long run.

7

Summary and Conclusion

The goal of this book was to develop an unsupervised speech controlled system which automatically adapts to several recurring users. They are allowed to operate the system without the requirement to attend a training. Instead, each user can directly use the speech controlled system without any constraints concerning vocabulary or the need to identify himself. Each utterance contains additional information about a particular speaker. The system can be incrementally personalized. Enrolled speakers have to be reliably identified to allow optimal speech decoding and continuous adjustment of the corresponding statistical models. New speaker profiles are initialized when unknown speakers are detected.

An implementation in an embedded system was intended and therefore computational complexity and memory consumption were essential design parameters. A compact representation with only one statistical model was targeted and an extension of a common speech recognizer was preferred.

The discussion started with speech production represented by a simple yet effective source-filter model. It defines the process of speech generation and provides a first overview of the complexity of speech and speaker variability and their dependencies. The acoustic environment, especially for in-car applications, is characterized by varying background noises and channel characteristics degrading automatic speech recognition and speaker identification.

Automated feature extraction was motivated by human hearing. The former one processes a continuous stream of audio data and extracts a feature vector representation for further pattern recognition. A standard technique for noise reduction was included.

Then several basic techniques for speaker change detection, speaker identification and speech recognition were discussed. They represent essential components of the intended solution. Step by step more sophisticated techniques and statistical models were introduced to capture speech and speaker characteristics.

Speaker adaptation moderates mismatches between training and test by adjusting the statistical models to unseen situations. Modeling accuracy

T. Herbig, F. Gerl, and W. Minker: Self-Learning Speaker Identification, SCT, pp. 145–148.
springerlink.com

depends on the number of parameters which can be reliably estimated. Several strategies were discussed suitable for fast and long-term adaptation depending on the amount of speaker specific training data.

Feature vector enhancement was introduced as an alternative approach to deal with speaker characteristics and environmental variations on a feature level.

Then some dedicated realizations of complete systems for unsupervised speaker identification or speech recognition were presented as found in the literature. Several strategies were considered to handle speaker variabilities and to establish more complex systems which enable an unsupervised speaker modeling and tracking. The solutions differ in model complexity and the mutual benefit between speech and speaker characteristics, for example. All these implementations cover several aspects of this book.

This was the starting point for the fusion of the basic techniques of speaker identification and speech recognition into one statistical model. A completely unsupervised system was developed:

First, an appropriate strategy for speaker adaptation was discussed. It is suitable for both short-term and long-term adaptation since the effective number of parameters are dynamically adjusted. Even on limited data a fast and robust retrieval of speaker related information is achieved. The speech recognizer is enabled to initialize and continuously adapt speaker specific models. One important property, especially for embedded devices, is the moderate complexity.

The experiments carried out showed the gain of this approach compared to strictly speaker independent implementations and speaker adaptation based on a few utterances without speaker identification. The error rate of the speech recognizer could be reduced by 25 % compared to the speaker independent baseline. With respect to the implementation of short-term adaptation a relative error rate reduction of 20 % was achieved.

Then a first realization of the target system was presented. Speaker specific codebooks can be used for speech recognition and speaker identification leading to a compact representation of speech and speaker characteristics. Multiple recognition steps are not required to generate a transcription and to estimate the speaker identity. All components comprising speaker identification, speech recognition and speaker adaptation were integrated into a speech controlled system.

In the experiments speech recognition accuracy could be increased by this technique to 88.20 %. For comparison, the corresponding upper limit was 88.90 % WA when the speaker is known. The speaker independent baseline only achieved 85.23 % WA. An optimum of 94.64 % identification rate could be observed.

In addition, a reference system was implemented which integrates a standard technique for speaker identification. All implementations obtained lower identification rates but similar speech recognition results in a realistic closed-set scenario. Furthermore, it was obviously more complicated to reliably detect unknown speakers compared to the unified approach of the target system. For both systems unknown speakers appeared to be an unsolved problem because of the unacceptably high error rates of missed and falsely detected out-of-set speakers.

Speaker identification was therefore extended by long-term speaker tracking. Instead of likelihoods, posterior probabilities can be used to reflect not only the match between model and observed data but also the training situation of each speaker specific model. Both weakly and extensively trained models are treated appropriately due to prior knowledge about the evolution of the system. Speaker tracking across several utterances was described by HMMs based on speaker specific codebooks considered on an utterance level. The forward-backward algorithm allows a more robust speaker identification since a series of utterances is evaluated. Furthermore, additional devices such as a beamformer can be easily integrated to support speaker change detection. No structural modifications of the speech controlled system are required. In addition, open-set scenarios can be handled. New codebooks can be initialized without imposing new users to identify themselves. No time-consuming enrollment is necessary.

The experiments for supervised adaptation and the first implementation of integrated speaker identification and speech recognition were repeated. Remarkable identification rates up to 98.59 % could be obtained in the closed-set scenario. Nearly the same speech recognition rate was observed. In an additional experiment the number of enrolled speakers was doubled. The identification rate on an utterance level decreased from 94.64 % to 89.56 %. However, the identification rate was raised again by speaker tracking so that 97.89 % of the enrolled speakers were correctly identified. Furthermore, speech recognition accuracy could be improved for some realizations.

Finally, the former experiment was considered in an open-set scenario. For speech recognition approximately 17 % and 12 % relative error rate reduction compared to the baseline and short-term adaptation were measured. Since the WA approached the results of the closed-set implementation in the long run, even higher gains might be expected for more extensive tests. For closed-sets a benefit for speech recognition could only be observed when long-term speaker tracking is applied to larger speaker sets. However, it is hard to imagine that unknown speakers can be reliably detected without long-term speaker tracking.

Potential for further improvements was seen in an enhanced feature extraction. If an optimal feature normalization can be guaranteed for the current speaker, both the speaker identification and speech recognition can benefit even on an utterance level. Under favorable conditions an optimum of 97.33 % correctly identified speakers and 88.61 % WA for speech recognition were obtained.

It can be concluded that a compact representation of speech and speaker characteristics was achieved which allows simultaneous speech recognition and speaker identification. Obviously, an additional module for speaker identification is not required. Only calculations of low complexity have to be performed in real-time since speaker tracking and adaptation are computed after utterance is finished. Therefore applications in embedded systems become feasible. The approach presented here has potential for further extensions and may be applied in a wide range of scenarios as sketched in the outlook.

8

Outlook

At the end of this book an outlook for prospective applications and extensions of the techniques presented here is discussed. The outlook is intended to emphasize the potential for many applications and shall motivate for further research.

Adaptation

One important aspect of the speech controlled system presented in this book is to capture speaker characteristics fast and accurately for robust speaker identification. This task was accomplished by EV adaptation. To obtain optimal performance of EV adaptation, training and test conditions have to be balanced. Especially for applications in varying environments, re-estimating the eigenvoices may be applied to improve on the eigenspace given by the PCA [Nguyen et al., 1999].

Furthermore, different adaptation schemes may be investigated with respect to joint speaker identification and speech recognition as possible future work. For example, MLLR adaptation which is widely used in speaker adaptation, e.g. Ferràs et al. [2007, 2008]; Gales and Woodland [1996]; Leggetter and Woodland [1995b], may be integrated into the system. For example, the subspace approach presented by Zhu et al. [2010] may be a promising candidate.

Usability

The discussion was focused on a speech controlled system triggered by a push-to-talk button to simplify speech segmentation. A more convenient human-computer communication may be realized by an unsupervised endpoint detection. Standard VAD techniques based on energy and fundamental frequency are expected to have limited capabilities in adverse environments, e.g. in automobiles.

T. Herbig, F. Gerl, and W. Minker: Self-Learning Speaker Identification, SCT, pp. 149–152.
springerlink.com

Model-based speech segmentation may be implemented by similar approaches as speaker identification. For instance, GMMs of moderate complexity can be used to model speech and non-speech events [Herbig et al., 2008; Lee et al., 2004; Oonishi et al., 2010; Tsai et al., 2003]. A decision logic similar to the implementation of speaker specific speech recognition may allow a simple decision whether speech is present.

Alternatively, this problem may be solved by an additional specialized codebook of the speech recognizer to represent different background scenarios during speech pauses. When a speech pause is detected, speech decoding can be rejected. More sophisticated solutions comprising Viterbi time alignment of all speaker models similar to the decoder-guided audio segmentation are also possible.

Barge-in functionality can be incorporated into speech segmentation to allow users to interrupt speech prompts. A more natural and efficient control of automated human-computer interfaces should be targeted. In the literature [Ittycheriah and Mammone, 1999; Ljolje and Goffin, 2007] similar techniques exist which might be integrated. Crosstalk detection may be an additional challenge [Song et al., 2010; Wrigley et al., 2005].

Development of the Unified Modeling of Speech and Speaker-Related Characteristics

The solution presented in this book is developed on a unified modeling of speech and speaker characteristics. The speech recognizer's codebooks are considered as common GMMs with uniform weights. Since the likelihood values are buffered until the speech recognition result is available, the phoneme alignment can be used not only to support speaker adaptation but also speaker identification in a similar way. Since an estimate for the state sequence is accessible, a refined likelihood becomes feasible. The restriction of equal weights can be avoided so that speaker identification can apply state dependent GMMs for each frame which may result in a better statistical representation.

Furthermore, some phoneme groups are well suited for speaker identification whereas others do not significantly contribute to speaker discrimination as outlined in Sect. 2.1 and Sect. 3.3. In combination with a phoneme alignment emphasis can be placed on vowels and nasals. The influence of fricatives and plosives on speaker identification can be reduced, for instance. Together this would result in a closer relationship between speech and speaker related information.

Enhancing Speaker Identification and Speech Recognition

Another method to increase the WA and identification rate is to employ a short supervised enrollment. Significant improvements are expected when the spoken text and the speaker identity are known for a couple of utterances.

The approach presented in this book also allows hybrid solutions. Each user can attend a supervised enrollment of variable duration to initialize a robust speaker profile. The speech recognizer is always run in an open-set mode to protect these profiles against unknown speakers. If a new speaker operates the device for the first time and decides not to attend an enrollment, a new profile can be temporarily initialized and applied subsequently.

In addition, the concepts of the speaker adaptation scheme and long-term speaker tracking presented here might be applied to the reference speaker identification. Even though no essential improvement is expected for the WA, an increase in the speaker identification rate might be achieved when both identification techniques are combined.

Mobile Applications

Even though only in-car applications were evaluated in this book, mobile applications can be considered as well. If a self-learning speaker identification is used for speech controlled mobile devices, one important aspect should be the problem of environmental variations.

The front-end of the system presented in this book reduces background noises and compensates for channel characteristics by noise reduction and cepstral mean subtraction. However, artifacts are still present and are learned by speaker adaptation. Especially for mobile applications, e.g. a Personal Digital Assistant (PDA), this can be undesirable. The focus should be set here on the variety of background conditions, e.g. in-car, office, public places and home. All scenarios have to be covered appropriately.

A simple realization would be to use several codebooks for each speaker and environment. Alternatively, feature vector normalization or enhancement may be combined with speaker adaptation. Effects from the acoustic background should be suppressed. In the optimal case only speaker characteristics are incorporated into the statistical modeling. The latter approach seems to be preferable since only one speaker specific codebook is needed and different training levels of domain specific codebooks are circumvented.

Thus, combining feature vector enhancement and adaptation may be advantageous when speaker characteristics and environmental influences can be separated. Since eigenvoices achieved good results for speaker adaptation, compensation on a feature vector level implemented by the eigen-environment approach could be a promising candidate. The identification of the current background scenario might help to apply more sophisticated compensation and speaker adaptation techniques.

Interaction

The use cases so far have involved identifying persons only by external means [Herbig et al., 2010c] or by an enrolling phase [Herbig et al., 2010d,e], or such as here where no external information was used [Herbig et al., 2011].

However, a wide range of human-computer communication may be of interest for realistic applications:

A dialog engine may give information about the temporal or semantic coherence of a user interaction. During an ongoing dialog a user change is not expected. Feedback of the speaker identity allows special habits and the experience of the user about the speech controlled device to be taken into consideration. Whereas beginners can be offered a comprehensive introduction or help concerning the handling of the device, advanced users can be pointed how to use the device more efficiently.

An acoustic preprocessing engine, e.g. in an infotainment or in-car communication system [Schmidt and Haulick, 2006], may use beamforming on a microphone array which can track the direction of the sound signal. For example, driver and co-driver may enter voice commands in turn. This soft information could be integrated to improve the robustness of speaker tracking. In addition, devices which are usually registered to a particular user, e.g. car keys, mobile phones or PDAs, can support speaker identification. Visual information may also be employed [Maragos et al., 2008].

In summary, a flexible framework has been developed which shows a high potential for future applications and research.

A

Appendix

A.1 Expectation Maximization Algorithm

The Expectation Maximization (EM) algorithm [Dempster et al., 1977] is a fundamental approach for mathematical optimization problems comprising a hidden or latent variable. An optimal statistical representation of a training data set $\mathbf{x}_{1:T}$ is targeted. A statistical model with parameter set Θ_{ML} characterized by

$$\Theta_{\mathrm{ML}} = \arg\max_{\Theta} \{p(\mathbf{x}_{1:T}|\Theta)\} \tag{A.1}$$

has to be determined. An iterative procedure is required due to the latent variable. In this section the meaning of the latent variable is explained for GMM training which was introduced in Sect. 2.4.2. Then the two steps of the EM algorithm are described.

Latent variables can be intuitively explained when a GMM is considered as a random generator. GMMs can be explained by a combination of two random processes as given by (2.35). For convenience, the first process is used to select the Gaussian density with index k according to the prior probability $p(k|\Theta)$. A second random generator produces a Gaussian distributed random vector $\mathbf{x}_t$ characterized by the probability density function $p(\mathbf{x}_t|k, \Theta)$. By repetition of these two steps, a sequence of feature vectors is produced. The assignment of the feature vector $\mathbf{x}_t$ to the corresponding Gaussian density $p(\mathbf{x}_t|k, \Theta)$ given by the first random generator is hidden for an observer. Therefore, the index k denotes the latent variable of GMM modeling.

GMM training or adaptation encounters the inverse problem since only the measured feature vectors $\mathbf{x}_{1:T}$ are available. The information which Gaussian density generated an observed feature vector is missing. Thus, training and adaptation algorithms require knowledge or at least an estimate concerning the corresponding Gaussian density.

Feature vectors $\mathbf{x}$ are subsequently called *incomplete data* since only the observations of the second random process are accessible. In contrast, *complete data* $(\mathbf{x}, k)$ comprise feature vector and latent variable.

Therefore, the likelihood of complete data $p(\mathbf{x}_t, k|\Theta)$ instead of incomplete data $p(\mathbf{x}_t|\Theta)$ is employed in the EM algorithm. To simplify the following considerations, the iid assumption is used. The problem of parameter optimization is solved in two steps:

- The *E-step* computes an estimate of the missing index k with the help of the posterior probability $p(k|\mathbf{x}_t, \bar{\Theta})$ based on initial parameters or the parameter set of the last iteration $\bar{\Theta}$. This posterior probability represents the responsibility of a Gaussian density for the production of the observation $\mathbf{x}_t$ and controls the effect on the training or adaptation of a particular Gaussian density.
- The *M-step* maximizes the likelihood function by calculating a new set of model parameters $\hat{\Theta}$ based on complete data given by the previous E-step.

E-step and M-step can be summarized for a sequence of training or adaptation data $\mathbf{x}_{1:T}$ in a compact mathematical description which is known as the auxiliary function

$$Q_{\mathrm{ML}}(\Theta, \bar{\Theta}) = \mathrm{E}_{\mathrm{k}_{1:T}}\{\log\left(p(\mathbf{x}_{1:T}, k_{1:T}|\Theta)\right)|\mathbf{x}_{1:T}, \bar{\Theta}\}. \tag{A.2}$$

It can be calculated by

$$Q_{\mathrm{ML}}(\Theta, \bar{\Theta}) = \sum_{t=1}^{T}\sum_{k=1}^{N} p(k|\mathbf{x}_t, \bar{\Theta}) \cdot \log\left(p(\mathbf{x}_t, k|\Theta)\right) \tag{A.3}$$

where the new parameter set $\hat{\Theta}$ is characterized by

$$\frac{\mathrm{d}}{\mathrm{d}\Theta} Q_{\mathrm{ML}}(\Theta, \bar{\Theta})\bigg|_{\Theta=\hat{\Theta}} = 0. \tag{A.4}$$

An iterative procedure is required since the assignment of the feature vectors $\mathbf{x}_t$ to a particular Gaussian density may be different for an updated parameter set $\hat{\Theta}$ compared to the preceding parameter set $\bar{\Theta}$. Several iterations of the EM algorithm are performed until an optimum of the likelihood $p(\mathbf{x}_{1:T}|\Theta)$ is obtained. However, local and global maxima cannot be distinguished.

For limited training data or in an adaptation scenario this might be inadequate. The risk of over-fitting arises since insignificant properties of the observed data are learned instead of general statistical patterns. The problem of over-fitting can be decreased when equation (A.3) is extended by prior knowledge $p(\Theta)$ about the parameter set Θ. The corresponding auxiliary function is then given by

$$Q_{\mathrm{MAP}}(\Theta, \bar{\Theta}) = Q_{\mathrm{ML}}(\Theta, \bar{\Theta}) + \log\left(p(\Theta)\right). \tag{A.5}$$

For further reading on the EM algorithm, GMM and HMM training the reader is referred to Bishop [2007]; Dempster et al. [1977]; Rabiner and Juang [1993].

A.2 Bayesian Adaptation

In this section codebook adaptation is exemplified by adapting the mean vectors in the Bayesian framework. The goal is an improved statistical representation of speech characteristics and speaker specific pronunciation for enhanced speech recognition. It is assumed that a set of speaker specific training data $\mathbf{x}_{1:T}$ is accessible.

A two-stage procedure is applied similar to the segmental MAP algorithm introduced in Sect. 2.6.2. Speech decoding provides an optimal state sequence and determines the state dependent weights w_k^s. Then the codebook of the current speaker i is adapted.

For convenience, the subsequent notation neglects the optimal state sequence to obtain a compact representation. In the following derivation, the auxiliary function of the EM algorithm

$$Q_{\mathrm{MAP}}(\Theta_i, \Theta_0) = Q_{\mathrm{ML}}(\Theta_i, \Theta_0) + \log\left(p(\Theta_i)\right) \tag{A.6}$$

$$= \sum_{t=1}^{T}\sum_{k=1}^{N} p(k|\mathbf{x}_t, \Theta_0) \cdot \log\left(p(\mathbf{x}_t, k|\Theta_i)\right) + \log\left(p(\Theta_i)\right) \tag{A.7}$$

is extended by a term comprising prior knowledge as given in (A.3) and (A.5). Only one iteration of the EM algorithm is calculated. The initial parameter set

$$\Theta_0 = \left\{w_1^0, \ldots, w_N^0, \boldsymbol{\mu}_1^0, \ldots, \boldsymbol{\mu}_N^0, \boldsymbol{\Sigma}_1^0, \ldots, \boldsymbol{\Sigma}_N^0\right\} \tag{A.8}$$

is given by the standard codebook. Since only mean vectors are adapted, the following notation is used for the speaker specific codebooks:

$$\Theta_i = \left\{w_1^0, \ldots, w_N^0, \boldsymbol{\mu}_1^i, \ldots, \boldsymbol{\mu}_N^i, \boldsymbol{\Sigma}_1^0, \ldots, \boldsymbol{\Sigma}_N^0\right\}. \tag{A.9}$$

Subsequently, the speaker index i is omitted. For reasons of simplicity, $\boldsymbol{\mu}_k$ and $\boldsymbol{\Sigma}_k$ denote the speaker specific mean vectors to be optimized and the covariance matrices of the standard codebook.

For the following equations it is assumed that each Gaussian distribution can be treated independently from the remaining distributions. Thus, the prior distribution of the speaker specific mean vector $\boldsymbol{\mu}_k$ can be factorized or equivalently the logarithm is given by a sum of logarithms. A prior Gaussian distribution is assumed

$$\log\left(p(\boldsymbol{\mu}_1, \ldots, \boldsymbol{\mu}_N)\right) = \sum_{k=1}^{N} \log\left(\mathcal{N}\left\{\boldsymbol{\mu}_k | \boldsymbol{\mu}_k^{\mathrm{pr}}, \boldsymbol{\Sigma}_k^{\mathrm{pr}}\right\}\right) \tag{A.10}$$

$$
\begin{aligned}
&= -\frac{1}{2}\sum_{k=1}^{N} (\boldsymbol{\mu}_k - \boldsymbol{\mu}_k^{\mathrm{pr}})^T \cdot (\boldsymbol{\Sigma}_k^{\mathrm{pr}})^{-1} \cdot (\boldsymbol{\mu}_k - \boldsymbol{\mu}_k^{\mathrm{pr}}) \\
&\quad - \frac{1}{2}\sum_{k=1}^{N} \log\left(|\boldsymbol{\Sigma}_k^{\mathrm{pr}}|\right) - \frac{d}{2}\sum_{k=1}^{N} \log(2\pi)
\end{aligned}
\tag{A.11}
$$

where the covariance matrix $\boldsymbol{\Sigma}_k^{\mathrm{pr}}$ represents the uncertainty of the adaptation. $\boldsymbol{\mu}_k^{\mathrm{pr}}$ may be given by the mean vectors of the standard codebook, for example.

Q_{MAP} is maximized by taking the derivative with respect to the mean vector $\boldsymbol{\mu}_k$ and by calculating the corresponding roots. Subsequently, equation (A.11) is inserted into (A.7) and the derivative

$$
\begin{aligned}
\frac{\mathrm{d}}{\mathrm{d}\boldsymbol{\mu}_k} Q_{\mathrm{MAP}}(\Theta, \Theta_0) &= \sum_{t=1}^{T} p(k|\mathbf{x}_t, \Theta_0) \cdot \frac{\mathrm{d}}{\mathrm{d}\boldsymbol{\mu}_k} \log\left(p(\mathbf{x}_t, k|\Theta)\right) \\
&\quad + \frac{\mathrm{d}}{\mathrm{d}\boldsymbol{\mu}_k} \log\left(p(\Theta)\right)
\end{aligned}
\tag{A.12}
$$

is calculated for a particular Gaussian density k.

The derivative

$$
\frac{\mathrm{d}}{\mathrm{d}\boldsymbol{\mu}_k} \log\left(p(\Theta)\right) = \frac{\mathrm{d}}{\mathrm{d}\boldsymbol{\mu}_k} \log\left(p(\boldsymbol{\mu}_1, \ldots, \boldsymbol{\mu}_N)\right) \tag{A.13}
$$

removes the constant terms in (A.11) comprising normalization and the determinant of the covariance matrix. Only the derivative of the squared Mahalanobis distance

$$
\bar{d}_k^{\mathrm{Mahal}} = (\boldsymbol{\mu}_k - \boldsymbol{\mu}_k^{\mathrm{pr}})^T \cdot (\boldsymbol{\Sigma}_k^{\mathrm{pr}})^{-1} \cdot (\boldsymbol{\mu}_k - \boldsymbol{\mu}_k^{\mathrm{pr}}) \tag{A.14}
$$

is retained in

$$
\begin{aligned}
\frac{\mathrm{d}}{\mathrm{d}\boldsymbol{\mu}_k} \log\left(p(\boldsymbol{\mu}_1, \ldots, \boldsymbol{\mu}_N)\right) &= -\frac{1}{2} \cdot \frac{\mathrm{d}}{\mathrm{d}\boldsymbol{\mu}_k} \bar{d}_k^{\mathrm{Mahal}} && \text{(A.15)} \\
&= -(\boldsymbol{\Sigma}_k^{\mathrm{pr}})^{-1} \cdot (\boldsymbol{\mu}_k - \boldsymbol{\mu}_k^{\mathrm{pr}}) && \text{(A.16)}
\end{aligned}
$$

according to (2.113).

Since the weights are given by the standard codebook, the derivative

$$
\begin{aligned}
\frac{\mathrm{d}}{\mathrm{d}\boldsymbol{\mu}_k} Q_{\mathrm{ML}}(\Theta, \Theta_0) &= \sum_{t=1}^{T} p(k|\mathbf{x}_t, \Theta_0) \cdot \frac{\mathrm{d}}{\mathrm{d}\boldsymbol{\mu}_k} \log\left(\mathcal{N}\{\mathbf{x}_t|\boldsymbol{\mu}_k, \boldsymbol{\Sigma}_k\}\right) && \text{(A.17)} \\
&= -\frac{1}{2}\sum_{t=1}^{T} p(k|\mathbf{x}_t, \Theta_0) \cdot \frac{\mathrm{d}}{\mathrm{d}\boldsymbol{\mu}_k} d_k^{\mathrm{Mahal}} && \text{(A.18)}
\end{aligned}
$$

only depends on the squared Mahalanobis distance

$$d_k^{\mathrm{Mahal}} = (\mathbf{x}_t - \boldsymbol{\mu}_k)^T \cdot \boldsymbol{\Sigma}_k^{-1} \cdot (\mathbf{x}_t - \boldsymbol{\mu}_k) \tag{A.19}$$

and the posterior probability $p(k|\mathbf{x}_t, \Theta_0)$. The derivative of Q_{ML} is given by

$$\frac{\mathrm{d}}{\mathrm{d}\boldsymbol{\mu}_k} Q_{\mathrm{ML}}(\Theta, \Theta_0) = \sum_{t=1}^{T} p(k|\mathbf{x}_t, \Theta_0) \cdot \boldsymbol{\Sigma}_k^{-1} \cdot (\mathbf{x}_t - \boldsymbol{\mu}_k) \,. \tag{A.20}$$

By using (A.20) and (A.16) in (A.12) a compact notation of the optimization problem can be obtained:

$$\begin{aligned} \frac{\mathrm{d}}{\mathrm{d}\boldsymbol{\mu}_k} Q_{\mathrm{MAP}}(\Theta, \Theta_0) = &\sum_{t=1}^{T} p(k|\mathbf{x}_t, \Theta_0) \cdot \boldsymbol{\Sigma}_k^{-1} \cdot (\mathbf{x}_t - \boldsymbol{\mu}_k) \\ &- (\boldsymbol{\Sigma}_k^{\mathrm{pr}})^{-1} \cdot (\boldsymbol{\mu}_k - \boldsymbol{\mu}_k^{\mathrm{pr}}) \,. \end{aligned} \tag{A.21}$$

The mean vector $\boldsymbol{\mu}_k$ and the covariance matrix $\boldsymbol{\Sigma}_k^{-1}$ of the standard codebook are time invariant and can be placed in front of the sum

$$\begin{aligned} \frac{\mathrm{d}}{\mathrm{d}\boldsymbol{\mu}_k} Q_{\mathrm{MAP}}(\Theta, \Theta_0) = &\boldsymbol{\Sigma}_k^{-1} \cdot \left(\sum_{t=1}^{T} p(k|\mathbf{x}_t, \Theta_0) \cdot \mathbf{x}_t - n_k \cdot \boldsymbol{\mu}_k \right) \\ &- (\boldsymbol{\Sigma}_k^{\mathrm{pr}})^{-1} \cdot (\boldsymbol{\mu}_k - \boldsymbol{\mu}_k^{\mathrm{pr}}) \end{aligned} \tag{A.22}$$

where

$$n_k = \sum_{t=1}^{T} p(k|\mathbf{x}_t, \Theta_0). \tag{A.23}$$

The normalized sum over all observed feature vectors weighted by the posterior probability is defined in (2.43) as the ML estimate

$$\boldsymbol{\mu}_k^{\mathrm{ML}} = \frac{1}{n_k} \sum_{t=1}^{T} p(k|\mathbf{x}_t, \Theta_0) \cdot \mathbf{x}_t. \tag{A.24}$$

The optimization problem

$$\left. \frac{\mathrm{d}}{\mathrm{d}\boldsymbol{\mu}_k} Q_{\mathrm{MAP}}(\Theta, \Theta_0) \right|_{\boldsymbol{\mu}_k = \boldsymbol{\mu}_k^{\mathrm{opt}}} = 0 \tag{A.25}$$

is solved by

$$n_k \cdot \boldsymbol{\Sigma}_k^{-1} \cdot (\boldsymbol{\mu}_k^{\mathrm{ML}} - \boldsymbol{\mu}_k^{\mathrm{opt}}) = (\boldsymbol{\Sigma}_k^{\mathrm{pr}})^{-1} \cdot (\boldsymbol{\mu}_k^{\mathrm{opt}} - \boldsymbol{\mu}_k^{\mathrm{pr}}) \tag{A.26}$$

which is known as Bayesian adaptation [Duda et al., 2001].

A.3 Evaluation Measures

Speech recognition is evaluated in this book by the so-called Word Accuracy (WA) which is widely accepted in the speech recognition literature [Boros et al., 1996]. The recognized word sequence is compared with a reference string to determine the word accuracy

$$\mathrm{WA} = 100 \cdot \left(1 - \frac{W_{\mathrm{S}} + W_{\mathrm{I}} + W_{\mathrm{D}}}{W}\right) \%. \tag{A.27}$$

W represents the number of words in the reference string and W_{S}, W_{I} and W_{D} denote the number of substituted, inserted and deleted words in the recognized string [Boros et al., 1996]. Alternatively, the *Word Error Rate* (WER) defined by

$$\mathrm{WER} = 100 \cdot \left(\frac{W_{\mathrm{S}} + W_{\mathrm{I}} + W_{\mathrm{D}}}{W}\right) \% \tag{A.28}$$

can be employed [Bisani and Ney, 2004].

Speaker identification is evaluated by the identification rate ζ which is approximated by the ratio

$$\hat{\zeta} = \frac{\sum_{u=1}^{N_u} \delta_{\mathrm{K}}(i_u^{\mathrm{MAP}}, i_u)}{N_u} \tag{A.29}$$

of correctly assigned utterances and the total number of utterances. In this book, only those utterances are investigated which lead to a speech recognition result.

In addition, a confidence interval is given for identification rates[1] to reflect the statistical uncertainty. The uncertainty is caused by the limited number of utterances incorporated into the estimation of the identification rate. Therefore, the interval is determined in which the true identification rate ζ has to be expected with a predetermined probability $1 - \alpha$ given an experimental mean estimate $\hat{\zeta}$ based on N_u utterances:

$$p(z_{1-\frac{\alpha}{2}} < \frac{(\hat{\zeta} - \zeta) \cdot \sqrt{N_u}}{\hat{\sigma}} \leq z_{\frac{\alpha}{2}}) = 1 - \alpha. \tag{A.30}$$

z is the cut-off of the normalized random variable and $\hat{\sigma}$ denotes the standard deviation of the applied estimator. Subsequently, the Gaussian approximation is used to compute the confidence interval

$$\hat{\zeta} - \frac{\hat{\sigma} \cdot t_{\frac{\alpha}{2}}}{\sqrt{N_u}} < \zeta \leq \hat{\zeta} + \frac{\hat{\sigma} \cdot t_{\frac{\alpha}{2}}}{\sqrt{N_u}} \tag{A.31}$$

[1] Even though the WA is a rate and not a probability [Bisani and Ney, 2004], equation (A.31) is equally employed in this book to indicate the reliability of the speech recognition rate.

of a Binomial distribution as found by Kerekes [2008]. The standard deviation $\hat{\sigma}$ is given by $\hat{\sigma}^2 = \hat{\zeta} \cdot (1 - \hat{\zeta})$. $t_{\frac{\alpha}{2}}$ is characterized by the Student distribution. In this context $\alpha = 0.05$ and $t_{\frac{\alpha}{2}} = 1.96$ are used. In the figures of this book confidence intervals are given by a gray shading.

Frequently, several realizations of an automated speech recognizer are evaluated and compared for certain values of a tuning parameter, e.g. in Table 4.1. Those tests are usually performed on identical data sets to avoid uncertainties caused by independent data. In this case paired difference tests are appropriate to determine whether a significant difference between two experiments can be assumed for a particular data set [Bisani and Ney, 2004; Bortz, 2005]. For example, the typical error of the results shown in Table 4.1 is about $\pm 0.2\,\%$ for independent datasets and about $\pm 0.03\,\%$ for this specific evaluation. In this book only the typical errors of independent datasets are given for reasons of simplicity.

The problem of speaker change detection is characterized by a binary decision problem. H_0 and H_1 denote the hypotheses of no speaker change and a speaker change, respectively. The optimal classifier is given in (2.14). The following two error scenarios determine the performance of the binary classifier:

$$\begin{aligned} &p(H_1 | i_{u-1} = i_u) \quad \text{false alarm,} \\ &p(H_0 | i_{u-1} \neq i_u) \quad \text{missed speaker change.} \end{aligned}$$

Alternatively, the accuracy of a binary classifier can be graphically represented for varying thresholds by the Receiver Operation Characteristics (ROC) curve[2]. The true positives (which are complimentary to the miss rate) are plotted versus the false alarm rate independently from the prior probability of a speaker change.

For details on confidence measures and ROC curves it is referred to Kerekes [2008]; Macskassy and Provost [2004].

[2] Detection Error Trade-off (DET) curves can be equivalently used to graphically represent both error types of a binary classifier [Martin et al., 1997].

References

Ajmera, J., McCowan, I., Bourlard, H.: Robust speaker change detection. IEEE Signal Processing Letters 11(8), 649–651 (2004)

Angkititrakul, P., Hansen, J.H.L.: Discriminative in-set/out-of-set speaker recognition. IEEE Transactions on Audio, Speech, and Language Processing 15(2), 498–508 (2007)

Barras, C., Gauvain, J.-L.: Feature and score normalization for speaker verification of cellular data. In: IEEE International Conference on Acoustics, Speech, and Signal Processing, ICASSP 2003, vol. 2, pp. 49–52 (2003)

Benesty, J., Makino, S., Chen, J. (eds.): Speech Enhancement. Signals and Communication Technology. Springer, Heidelberg (2005)

Bennett, P.N.: Using asymmetric distributions to improve text classifier probability estimates. In: Proceedings of the 26th Annual International ACM SIGIR Conference on Research and Development in Information Retrieval, SIGIR 2003, pp. 111–118 (2003)

Bimbot, F., Magrin-Chagnolleau, I., Mathan, L.: Second-order statistical measures for text-independent speaker identification. Speech Communication 17(1-2), 177–192 (1995)

Bimbot, F., Mathan, L.: Text-free speaker recognition using an arithmetic-harmonic sphericity measure. In: EUROSPEECH 1993, pp. 169–172 (1993)

Bisani, M., Ney, H.: Bootstrap estimates for confidence intervals in ASR performance evaluation. In: IEEE International Conference on Acoustics, Speech, and Signal Processing, ICASSP 2004, vol. 1, pp. 409–412 (2004)

Bishop, C.M.: Neural Networks for Pattern Recognition, 1st edn. Oxford University Press, Oxford (1996)

Bishop, C.M.: Pattern Recognition and Machine Learning, 1st edn. Springer, New York (2007)

Boros, M., Eckert, W., Gallwitz, F., Gorz, G., Hanrieder, G., Niemann, H.: Towards understanding spontaneous speech: word accuracy vs. concept accuracy vs. concept accuracy. In: International Conference on Spoken Language Processing, ICSLP 1996, vol. 2, pp. 1009–1012 (1996)

Bortz, J.: Statistik: Für Human- und Sozialwissenschaftler, 6th edn. Springer-Lehrbuch. Springer, Heidelberg (2005) (in German)

Botterweck, H.: Anisotropic MAP defined by eigenvoices for large vocabulary continuous speech recognition. In: IEEE International Conference on Acoustics, Speech, and Signal Processing, ICASSP 2001, vol. 1, pp. 353–356 (2001)

Britanak, V., Yip, P., Rao, K.R.: Discrete Cosine and Sine Transforms: General Properties, Fast Algorithms and Integer Approximations. Academic Press, London (2006)

Bronstein, I.N., Semendjajew, K.A., Musiol, G., Muehlig, H.: Taschenbuch der Mathematik. Verlag Harri Deutsch, Frankfurt am Main (2000) (in German)

Buera, L., Lleida, E., Miguel, A., Ortega, A., Saz, Ó.: Cepstral vector normalization based on stereo data for robust speech recognition. IEEE Transactions on Audio, Speech and Language Processing 15(3), 1098–1113 (2007)

Buera, L., Lleida, E., Rosas, J.D., Villalba, J., Miguel, A., Ortega, A., Saz, O.: Speaker verification and identification using phoneme dependent multi-environment models based linear normalization in adverse and dynamic acoustic environments. In: Summer School for Advanced Studies on Biometrics for Secure Authentication: Multimodality and System Integration (2005)

Campbell, J.P.: Speaker recognition - a tutorial. Proceedings of the IEEE 85(9), 1437–1462 (1997)

Cappé, O.: Elimination of the musical noise phenomenon with the Ephraim and Malah noise suppressor. IEEE Transactions on Speech and Audio Processing 2(2), 345–349 (1994)

Che, C., Lin, Q., Yuk, D.-S.: An HMM approach to text-prompted speaker verification. In: IEEE International Conference on Acoustics, Speech, and Signal Processing, ICASSP 1996, vol. 2, pp. 673–676 (1996)

Chen, S.S., Gopalakrishnan, P.S.: Speaker, environment and channel change detection and clustering via the Bayesian information criterion. In: Proceedings of the DARPA Broadcast News Transcription and Understanding Workshop, pp. 127–132 (1998)

Cheng, S.-S., Wang, H.-M.: A sequential metric-based audio segmentation method via the Bayesian information criterion. In: EUROSPEECH 2003, pp. 945–948 (2003)

Class, F., Haiber, U., Kaltenmeier, A.: Automatic detection of change in speaker in speaker adaptive speech recognition systems. US Patent Application 2003/0187645 A1 (2003)

Class, F., Kaltenmeier, A., Regel-Brietzmann, P.: Optimization of an HMM - based continuous speech recognizer. In: EUROSPEECH 1993, pp. 803–806 (1993)

Class, F., Kaltenmeier, A., Regel-Brietzmann, P.: Speech recognition method with adaptation of the speech characteristics. European Patent EP0586996 (1994)

Cohen, I.: Noise spectrum estimation in adverse environments: Improved minima controlled recursive averaging. IEEE Transactions on Speech and Audio Processing 11(5), 466–475 (2003)

Cohen, I., Berdugo, B.: Noise estimation by minima controlled recursive. IEEE Signal Processing Letters 9(1), 12–15 (2002)

Davis, S.B., Mermelstein, P.: Comparison of parametric representations for monosyllabic word recognition in continuously spoken sentences. IEEE Transactions on Acoustics, Speech, and Signal Processing 28(4), 357–366 (1980)

Delakis, M., Gravier, G., Gros, P.: Stochastic models for multimodal video analysis. In: Maragos, P., Potamianos, A., Gros, P. (eds.) Multimodal Processing and Interaction: Audio, Video, Text, 1st edn. Multimedia Systems and Applications, vol. 33, pp. 91–109. Springer, New York (2008)

Dempster, A.P., Laird, N.M., Rubin, D.B.: Maximum likelihood from incomplete data via the EM algorithm. Journal of the Royal Statistical Society Series B 39(1), 1–38 (1977)
Dobler, S., Rühl, H.-W.: Speaker adaptation for telephone based speech dialogue systems. In: EUROSPEECH 1995, pp. 1139–1143 (1995)
Droppo, J., Acero, A.: Maximum mutual information SPLICE transform for seen and unseen conditions. In: INTERSPEECH 2005, pp. 989–992 (2005)
Droppo, J., Deng, L., Acero, A.: Evaluation of the SPLICE algorithm on the Aurora2 database. In: EUROSPEECH 2001, pp. 217–220 (2001)
Duda, R.O., Hart, P.E., Stork, D.G.: Pattern Classification, 2nd edn. Wiley Interscience, New York (2001)
Eatock, J.P., Mason, J.S.: A quantitative assessment of the relative speaker discriminating properties of phonemes. In: IEEE International Conference on Acoustics, Speech, and Signal Processing, ICASSP 1994, vol. 1, pp. 133–136 (1994)
Ephraim, Y., Malah, D.: Speech enhancement using a minimum mean-square error short-time spectral amplitude estimator. IEEE Transactions on Acoustics, Speech, and Signal Processing ASSP-32(6), 1109–1121 (1984)
Espi, M., Miyabe, S., Nishimoto, T., Ono, N., Sagayama, S.: Analysis on speech characteristics for robust voice activity detection. In: IEEE Workshop on Spoken Language Technology, SLT 2010, pp. 139–144 (2010)
Faltlhauser, R., Ruske, G.: Improving speaker recognition using phonetically structured Gaussian mixture models. In: EUROSPEECH 2001, pp. 751–754 (2001)
Fant, G.: Acoustic Theory of Speech Production. Mouton de Gruyter, The Hague (1960)
Felippa, C.A.: Introduction to finite element methods (2004)
Ferràs, M., Leung, C.C., Barras, C., Gauvain, J.-L.: Constrained MLLR for speaker recognition. In: IEEE International Conference on Acoustics, Speech, and Signal Processing, ICASSP 2007, vol. 4, pp. 53–56 (2007)
Ferràs, M., Leung, C.C., Barras, C., Gauvain, J.-L.: MLLR techniques for speaker recognition. In: The Speaker and Language Recognition Workshop, Odyssey 2008, pp. 21–24 (2008)
Fink, G.A.: Mustererkennung mit Markov-Modellen: Theorie-Praxis-Anwendungsgebiete. Leitfäden der Informatik. B. G. Teubner, Stuttgart (2003) (in German)
Forney, G.D.: The Viterbi algorithm. Proceedings of the IEEE 61(3), 268–278 (1973)
Fortuna, J., Sivakumaran, P., Ariyaeeinia, A., Malegaonkar, A.: Open-set speaker identification using adapted Gaussian mixture models. In: INTERSPEECH 2005, pp. 1997–2000 (2005)
Furui, S.: Selected topics from 40 years of research in speech and speaker recognition. In: INTERSPEECH 2009, pp. 1–8 (2009)
Gales, M.J.F., Woodland, P.C.: Mean and variance adaptation within the MLLR framework. Computer Speech and Language 10, 249–264 (1996)
Garau, G., Renals, S., Hain, T.: Applying vocal tract length normalization to meeting recordings. In: INTERSPEECH 2005, pp. 265–268 (2005)
Gauvain, J.-L., Lamel, L.F., Prouts, B.: Experiments with speaker verification over the telephone. In: EUROSPEECH 1995, pp. 651–654 (1995)
Gauvain, J.-L., Lee, C.-H.: Maximum a posteriori estimation for multivariate Gaussian mixture observations of Markov chains. IEEE Transactions on Speech and Audio Processing 2(2), 291–298 (1994)

Geiger, J., Wallhoff, F., Rigoll, G.: GMM-UBM based open-set online speaker diarization. In: INTERSPEECH 2010, pp. 2330–2333 (2010)
Genoud, D., Ellis, D., Morgan, N.: Combined speech and speaker recognition with speaker-adapted connectionist models. In: IEEE Workshop on Automatic Speech Recognition and Understanding, ASRU 1999 (1999)
Gish, H., Schmidt, M.: Text-independent speaker identification. IEEE Signal Processing Magazine 11(4), 18–32 (1994)
Godin, K.W., Hansen, J.H.L.: Session variability contrasts in the MARP corpus. In: INTERSPEECH 2010, pp. 298–301 (2010)
Goldenberg, R., Cohen, A., Shallom, I.: The Lombard effect's influence on automatic speaker verification systems and methods for its compensation. In: International Conference on Information Technology: Research and Education, ITRE 2006, pp. 233–237 (2006)
Gollan, C., Bacchiani, M.: Confidence scores for acoustic model adaptation. In: IEEE International Conference on Acoustics, Speech, and Signal Processing, ICASSP 2008, pp. 4289–4292 (2008)
Gutman, D., Bistritz, Y.: Speaker verification using phoneme-adapted Gaussian mixture models. In: The XI European Signal Processing Conference, EUSIPCO 2002, vol. 3, pp. 85–88 (2002)
Häb-Umbach, R.: Investigations on inter-speaker variability in the feature space. In: IEEE International Conference on Acoustics, Speech, and Signal Processing, ICASSP 1999, vol. 1, pp. 397–400 (1999)
Hain, T., Johnson, S.E., Tuerk, A., Woodland, P.C., Young, S.J.: Segment generation and clustering in the HTK broadcast news transcription system. In: Proceedings of the Broadcast News Transcription and Understanding Workshop, pp. 133–137 (1998)
Hänsler, E.: Statistische Signale: Grundlagen und Anwendungen, 3rd edn. Springer, Heidelberg (2001) (in German)
Hänsler, E., Schmidt, G.: Acoustic Echo and Noise Control: A Practical Approach. Wiley-IEEE Press, Hoboken (2004)
Hänsler, E., Schmidt, G. (eds.): Speech and Audio Processing in Adverse Environments. Signals and Communication Technology. Springer, Heidelberg (2008)
Harrag, A., Mohamadi, T., Serignat, J.F.: LDA combination of pitch and MFCC features in speaker recognition. In: IEEE Indicon Conference, pp. 237–240 (2005)
Hautamäki, V., Kinnunen, T., Nosratighods, M., Lee, K.A., Ma, B., Li, H.: Approaching human listener accuracy with modern speaker verification. In: INTERSPEECH 2010, pp. 1473–1476 (2010)
Herbig, T., Gaupp, O., Gerl, F.: Modellbasierte Sprachsegmentierung. In: 34. Jahrestagung für Akustik, DAGA 2008, pp. 259–260 (2008) (in German)
Herbig, T., Gerl, F.: Joint speaker identification and speech recognition for speech controlled applications in an automotive environment. In: International Conference on Acoustics, NAG-DAGA 2009, pp. 74–77 (2009)
Herbig, T., Gerl, F., Minker, W.: Detection of unknown speakers in an unsupervised speech controlled system. In: Lee, G.G., Mariani, J., Minker, W., Nakamura, S. (eds.) IWSDS 2010. LNCS, vol. 6392, pp. 25–35. Springer, Heidelberg (2010a)
Herbig, T., Gerl, F., Minker, W.: Evaluation of two approaches for speaker specific speech recognition. In: Lee, G.G., Mariani, J., Minker, W., Nakamura, S. (eds.) IWSDS 2010. LNCS, vol. 6392, pp. 36–47. Springer, Heidelberg (2010b)

Herbig, T., Gerl, F., Minker, W.: Fast adaptation of speech and speaker characteristics for enhanced speech recognition in adverse intelligent environments. In: The 6th International Conference on Intelligent Environments, IE 2010, pp. 100–105 (2010c)

Herbig, T., Gerl, F., Minker, W.: Simultaneous speech recognition and speaker identification. In: IEEE Workshop on Spoken Language Technology, SLT 2010, pp. 206–210 (2010d)

Herbig, T., Gerl, F., Minker, W.: Speaker tracking in an unsupervised speech controlled system. In: INTERSPEECH 2010, pp. 2666–2669 (2010e)

Herbig, T., Gerl, F., Minker, W.: Evolution of an adaptive unsupervised speech controlled system. In: IEEE Workshop on Evolving and Adaptive Intelligent Systems, EAIS 2011, pp. 163–169 (2011)

Huang, C.-H., Chien, J.-T., Wang, H.-M.: A new eigenvoice approach to speaker adaptation. In: The 4th International Symposium on Chinese Spoken Language Processing, ISCSLP 2004, pp. 109–112 (2004)

Huang, X.D., Jack, M.A.: Unified techniques for vector quantization and hidden Markov modeling using semi-continuous models. In: IEEE International Conference on Acoustics, Speech, and Signal Processing, ICASSP 1989, vol. 1, pp. 639–642 (1989)

Iskra, D., Grosskopf, B., Marasek, K., van den Heuvel, H., Diehl, F., Kiessling, A.: SPEECON - speech databases for consumer devices: Database specification and validation. In: Proceedings of the Third International Conference on Language Resources and Evaluation, LREC 2002, pp. 329–333 (2002)

Ittycheriah, A., Mammone, R.J.: Detecting user speech in barge-in over prompts using speaker identification methods. In: EUROSPEECH 1999, pp. 327–330 (1999)

Johnson, S.E.: Who spoke when? - Automatic segmentation and clustering for determining speaker turns. In: EUROSPEECH 1999, vol. 5, pp. 2211–2214 (1999)

Jolliffe, I.T.: Principal Component Analysis, 2nd edn. Springer Series in Statistics. Springer, Heidelberg (2002)

Jon, E., Kim, D.K., Kim, N.S.: EMAP-based speaker adaptation with robust correlation estimation. In: IEEE International Conference on Acoustics, Speech, and Signal Processing, ICASSP 2001, vol. 1, pp. 321–324 (2001)

Junqua, J.-C.: The influence of acoustics on speech production: A noise-induced stress phenomenon known as the Lombard reflex. Speech Communication 20(1-2), 13–22 (1996)

Junqua, J.-C.: Robust Speech Recognition in Embedded Systems and PC Applications. Kluwer Academic Publishers, Dordrecht (2000)

Kammeyer, K.D., Kroschel, K.: Digitale Signalverarbeitung, 4th edn. B. G. Teubner, Stuttgart (1998) (in German)

Kerekes, J.: Receiver operating characteristic curve confidence intervals and regions. IEEE Geoscience and Remote Sensing Letters 5(2), 251–255 (2008)

Kim, D.Y., Umesh, S., Gales, M.J.F., Hain, T., Woodland, P.C.: Using VTLN for broadcast news transcription. In: INTERSPEECH 2004, pp. 1953–1956 (2004)

Kimball, O., Schmidt, M., Gish, H., Waterman, J.: Speaker verification with limited enrollment data. In: EUROSPEECH 1997, pp. 967–970 (1997)

Kinnunen, T.: Designing a speaker-discriminative adaptive filter bank for speaker recognition. In: International Conference on Spoken Language Processing, ICSLP 2002, pp. 2325–2328 (2002)

Kinnunen, T.: Spectral Features for Automatic Text-Independent Speaker Recognition. Ph.lic. thesis, Department of Computer Science, University of Joensuu, Finland (2003)

Krim, H., Viberg, M.: Two decades of array signal processing research: the parametric approach. IEEE Signal Processing Magazine 13(4), 67–94 (1996)

Kuhn, R., Junqua, J.-C., Nguyen, P., Niedzielski, N.: Rapid speaker adaptation in eigenvoice space. IEEE Transactions on Speech and Audio Processing 8(6), 695–707 (2000)

Kuhn, R., Nguyen, P., Junqua, J.-C., Boman, R., Niedzielski, N., Fincke, S., Field, K., Contolini, M.: Fast speaker adaptation using a priori knowledge. In: IEEE International Conference on Acoustics, Speech, and Signal Processing, ICASSP 1999, vol. 2, pp. 749–752 (1999)

Kuhn, R., Nguyen, P., Junqua, J.-C., Goldwasser, L., Niedzielski, N., Fincke, S., Field, K., Contolini, M.: Eigenvoices for speaker adaptation. In: International Conference on Spoken Language Processing, ICSLP 1998, vol. 5, pp. 1771–1774 (1998)

Kuhn, R., Perronnin, F., Junqua, J.-C.: Time is money: Why very rapid adaptation matters. In: Adaptation 2001, pp. 33–36 (2001)

Kwon, S., Narayanan, S.: Unsupervised speaker indexing using generic models. IEEE Transactions on Speech and Audio Processing 13(5), 1004–1013 (2005)

Kwon, S., Narayanan, S.S.: Speaker change detection using a new weighted distance measure. In: International Conference on Spoken Language Processing, ICSLP 2002, pp. 2537–2540 (2002)

Lee, A., Nakamura, K., Nisimura, R., Saruwatari, H., Shikano, K.: Noise robust real world spoken dialogue system using GMM based rejection of unintended inputs. In: INTERSPEECH 2004, pp. 173–176 (2004)

Leggetter, C.J., Woodland, P.C.: Flexible speaker adaptation using maximum likelihood linear regression. In: Proceedings of the ARPA Spoken Language Technology Workshop, pp. 110–115 (1995a)

Leggetter, C.J., Woodland, P.C.: Maximum likelihood linear regression for speaker adaptation of continuous density hidden Markov models. Computer Speech and Language 9, 171–185 (1995b)

Lehn, J., Wegmann, H.: Einführung in die Statistik, 3rd edn. B. G. Teubner, Stuttgart (2000) (in German)

Linde, Y., Buzo, A., Gray, R.M.: An algorithm for vector quantizer design. IEEE Transactions on Communications 28(1), 84–95 (1980)

Liu, L., He, J., Palm, G.: A comparison of human and machine in speaker recognition. In: EUROSPEECH 1997, pp. 2327–2330 (1997)

Ljolje, A., Goffin, V.: Discriminative training of multi-state barge-in models. In: IEEE Workshop on Automatic Speech Recognition and Understanding, ASRU 2007, pp. 353–358 (2007)

Logan, B.: Mel frequency cepstral coefficients for music modeling. In: 1st International Symposium on Music Information Retrieval, ISMIR 2000 (2000)

Lu, L., Zhang, H.J.: Speaker change detection and tracking in real-time news broadcasting analysis. In: Proceedings of the Tenth ACM International Conference on Multimedia, MULTIMEDIA 2002, pp. 602–610 (2002)

Macskassy, S.A., Provost, F.J.: Confidence bands for ROC curves: Methods and an empirical study. In: 1st Workshop on ROC Analysis in Artificial Intelligence, ROCAI 2004, pp. 61–70 (2004)

Malegaonkar, A.S., Ariyaeeinia, A.M., Sivakumaran, P.: Efficient speaker change detection using adapted Gaussian mixture models. IEEE Transactions on Audio, Speech, and Language Processing 15(6), 1859–1869 (2007)

Mammone, R.J., Zhang, X., Ramachandran, R.P.: Robust speaker recognition. IEEE Signal Processing Magazine 13(5), 58–71 (1996)

Maragos, P., Potamianos, A., Gros, P. (eds.): Multimodal Processing and Interaction: Audio, Video, Text, 1st edn. Multimedia Systems and Applications, vol. 33. Springer, New York (2008)

Markov, K., Nakagawa, S.: Frame level likelihood normalization for text-independent speaker identification using Gaussian mixture models. In: International Conference on Spoken Language Processing, ICSLP 1996, vol. 3, pp. 1764–1767 (1996)

Martin, A., Doddington, G., Kamm, T., Ordowski, M., Przybocki, M.: The DET curve in assessment of detection task performance. In: EUROSPEECH 1997, pp. 1895–1898 (1997)

Matrouf, D., Bellot, O., Nocera, P., Linares, G., Bonastre, J.-F.: A posteriori and a priori transformations for speaker adaptation in large vocabulary speech recognition systems. In: EUROSPEECH 2001, pp. 1245–1248 (2001)

Matsui, T., Furui, S.: Concatenated phoneme models for text-variable speaker recognition. In: IEEE International Conference on Acoustics, Speech, and Signal Processing, ICASSP 1993, vol. 2, pp. 391–394 (1993)

Meinedo, H., Neto, J.: Audio segmentation, classification and clustering in a broadcast news task. In: IEEE International Conference on Acoustics, Speech, and Signal Processing, ICASSP 2003, vol. 2, pp. 5–8 (2003)

Meyberg, K., Vachenauer, P.: Höhere Mathematik 1: Differential- und Integralrechnung. Vektor- und Matrizenrechnung, 6th edn. Springer-Lehrbuch. Springer, Heidelberg (2003) (in German)

Meyer, B., Wesker, T., Brand, T., Mertins, A., Kollmeier, B.: A human-machine comparison in speech recognition based on a logatome corpus. In: Speech Recognition and Intrinsic Variation, SRIV 2006, pp. 95–100 (2006)

Moreno, P.J.: Speech Recognition in Noisy Environments. PhD thesis, Department of Electrical and Computer Engineering, Carnegie Mellon University, Pittsburgh, Pennsylvania (1996)

Mori, K., Nakagawa, S.: Speaker change detection and speaker clustering using VQ distortion for broadcast news speech recognition. In: IEEE International Conference on Acoustics, Speech, and Signal Processing, ICASSP 2001, vol. 1, pp. 413–416 (2001)

Morris, A.C., Wu, D., Koreman, J.: MLP trained to separate problem speakers provides improved features for speaker identification. In: 39th Annual International Carnahan Conference on Security Technology, CCST 2005, pp. 325–328 (2005)

Nakagawa, S., Zhang, W., Takahashi, M.: Text-independent / text-prompted speaker recognition by combining speaker-specific GMM with speaker adapted syllable-based HMM. IEICE Transactions on Information and Systems E89-D(3), 1058–1065 (2006)

Nguyen, P., Wellekens, C., Junqua, J.-C.: Maximum likelihood eigenspace and MLLR for speech recognition in noisy environments. In: EUROSPEECH 1999, pp. 2519–2522 (1999)

Nishida, M., Kawahara, T.: Speaker indexing and adaptation using speaker clustering based on statistical model selection. In: IEEE International Conference on Acoustics, Speech, and Signal Processing, ICASSP 2004, vol. 1, pp. 353–356 (2004)

Nishida, M., Kawahara, T.: Speaker model selection based on the Bayesian information criterion applied to unsupervised speaker indexing. IEEE Transactions on Speech and Audio Processing 13(4), 583–592 (2005)

Olsen, J.Ø.: Speaker verification based on phonetic decision making. In: EUROSPEECH 1997, pp. 1375–1378 (1997)

Oonishi, T., Iwano, K., Furui, S.: VAD-measure-embedded decoder with online model adaptation. In: INTERSPEECH 2010, pp. 3122–3125 (2010)

Oppenheim, A.V., Schafer, R.W.: Digital Signal Processing. Prentice-Hall, Englewood Cliffs (1975)

O'Shaughnessy, D.: Speech Communications: Human and Machine, 2nd edn. IEEE Press, New York (2000)

Park, A., Hazen, T.J.: ASR dependent techniques for speaker identification. In: International Conference on Spoken Language Processing, ICSLP 2002, pp. 1337–1340 (2002)

Pelecanos, J., Slomka, S., Sridharan, S.: Enhancing automatic speaker identification using phoneme clustering and frame based parameter and frame size selection. In: Proceedings of the Fifth International Symposium on Signal Processing and Its Applications, ISSPA 1999, vol. 2, pp. 633–636 (1999)

Petersen, K.B., Pedersen, M.S.: The matrix cookbook (2008)

Quatieri, T.F.: Discrete-Time Speech Signal Processing: Principles and Practice. Prentice-Hall, Upper Saddle River (2002)

Rabiner, L., Juang, B.-H.: Fundamentals of Speech Recognition. Prentice-Hall, Englewood Cliffs (1993)

Rabiner, L.R.: A tutorial on hidden Markov models and selected applications in speech recognition. Proceedings of the IEEE 77(2), 257–286 (1989)

Ramírez, J., Górriz, J.M., Segura, J.C.: Voice activity detection. Fundamentals and speech recognition system robustness. In: Grimm, M., Kroschel, K. (eds.) Robust Speech Recognition and Understanding, pp. 1–22. I-Tech Education and Publishing, Vienna (2007)

Reynolds, D., Campbell, J., Campbell, B., Dunn, B., Gleason, T., Jones, D., Quatieri, T., Quillen, C., Sturim, D., Torres-Carrasquillo, P.: Beyond cepstra: Exploiting high-level information in speaker recognition. In: Proceedings of the Workshop on Multimodal User Authentication, pp. 223–229 (2003)

Reynolds, D.A.: Large population speaker identification using clean and telephone speech. IEEE Signal Processing Letters 2(3), 46–48 (1995a)

Reynolds, D.A.: Speaker identification and verification using Gaussian mixture speaker models. Speech Communication 17(1-2), 91–108 (1995b)

Reynolds, D.A., Carlson, B.A.: Text-dependent speaker verification using decoupled and integrated speaker and speech recognizers. In: EUROSPEECH 1995, pp. 647–650 (1995)

Reynolds, D.A., Quatieri, T.F., Dunn, R.B.: Speaker verification using adapted Gaussian mixture models. Digital Signal Processing 10(1-3), 19–41 (2000)

Reynolds, D.A., Rose, R.C.: Robust text-independent speaker identification using Gaussian mixture speaker models. IEEE Transactions on Speech and Audio Processing 3(1), 72–83 (1995)

Rieck, S., Schukat-Talamazzini, E.G., Niemann, H.: Speaker adaptation using semi-continuous hidden Markov models. In: 11th IAPR International Conference on Pattern Recognition, vol. 3, pp. 541–544 (1992)
Rodríguez-Liñares, L., García-Mateo, C.: On the use of acoustic segmentation in speaker identification. In: EUROSPEECH 1997, pp. 2315–2318 (1997)
Rozzi, W.A., Stern, R.M.: Speaker adaptation in continuous speech recognition via estimation of correlated mean vectors. In: IEEE International Conference on Acoustics, Speech, and Signal Processing, ICASSP 1991, vol. 2, pp. 865–868 (1991)
Schmalenstroeer, J., Häb-Umbach, R.: Joint speaker segmentation, localization and identification for streaming audio. In: INTERSPEECH 2007, pp. 570–573 (2007)
Schmidt, G., Haulick, T.: Signal processing for in-car communication systems. Signal processing 86(6), 1307–1326 (2006)
Schukat-Talamazzini, E.G.: Automatische Spracherkennung. Vieweg, Braunschweig (1995) (in German)
Segura, J.C., Benítez, C., de la Torre, Á., Rubio, A.J., Ramírez, J.: Cepstral domain segmental nonlinear feature transformations for robust speech recognition. IEEE Signal Processing Letters 11(5), 517–520 (2004)
Setiawan, P., Höge, H., Fingscheidt, T.: Entropy-based feature analysis for speech recognition. In: INTERSPEECH 2009, pp. 2959–2962 (2009)
Siegler, M.A., Jain, U., Raj, B., Stern, R.M.: Automatic segmentation, classification and clustering of broadcast news audio. In: Proceedings of the DARPA Speech Recognition Workshop, pp. 97–99 (1997)
Song, H.J., Kim, H.S.: Eigen-environment based noise compensation method for robust speech recognition. In: INTERSPEECH 2005, pp. 981–984 (2005)
Song, J.-H., Lee, K.-H., Park, Y.-S., Kang, S.-I., Chang, J.-H.: On using Gaussian mixture model for double-talk detection in acoustic echo suppression. In: INTERSPEECH 2010, pp. 2778–2781 (2010)
Stern, R.M., Lasry, M.J.: Dynamic speaker adaptation for feature-based isolated word recognition. IEEE Transactions on Acoustics, Speech and Signal Processing 35(6), 751–763 (1987)
Thelen, E.: Long term on-line speaker adaptation for large vocabulary dictation. In: International Conference on Spoken Language Processing, ICSLP 1996, pp. 2139–2142 (1996)
Thompson, J., Mason, J.S.: Cepstral statistics within phonetic subgroups. In: International Conference on Signal Processing, ICSP 1993, pp. 737–740 (1993)
Thyes, O., Kuhn, R., Nguyen, P., Junqua, J.-C.: Speaker identification and verification using eigenvoices. In: International Conference on Spoken Language Processing, ICSLP 2000, vol. 2, pp. 242–245 (2000)
Tritschler, A., Gopinath, R.: Improved speaker segmentation and segments clustering using the Bayesian information criterion. In: EUROSPEECH 1999, vol. 2, pp. 679–682 (1999)
Tsai, W.-H., Wang, H.-M., Rodgers, D.: Automatic singer identification of popular music recordings via estimation and modeling of solo vocal signal. In: EUROSPEECH 2003, pp. 2993–2996 (2003)
Vary, P., Heute, U., Hess, W.: Digitale Sprachsignalverarbeitung. B.G. Teubner, Stuttgart (1998) (in German)
Veen, B.D.V., Buckley, K.M.: Beamforming: A versatile approach to spatial filtering. IEEE ASSP Magazine 5(2), 4–24 (1988)

Wendemuth, A.: Grundlagen der stochastischen Sprachverarbeitung. Oldenbourg (2004) (in German)

Wilcox, L., Kimber, D., Chen, F.: Audio indexing using speaker identification. In: Proceedings of the SPIE Conference on Automatic Systems for the Inspection and Identification of Humans, vol. 2277, pp. 149–157 (1994)

Wrigley, S.N., Brown, G.J., Wan, V., Renals, S.: Speech and crosstalk detection in multi-channel audio. IEEE Transactions on Speech and Audio Processing 13(1), 84–91 (2005)

Wu, T., Lu, L., Chen, K., Zhang, H.-J.: UBM-based real-time speaker segmentation for broadcasting news. In: IEEE International Conference on Acoustics, Speech, and Signal Processing, ICASSP 2003, vol. 2, pp. 193–196.

Yella, S.H., Varma, V., Prahallad, K.: Significance of anchor speaker segments for constructing extractive audio summaries of broadcast news. In: IEEE Workshop on Spoken Language Technology, SLT 2010, pp. 13–18 (2010)

Yin, S.-C., Rose, R., Kenny, P.: Adaptive score normalization for progressive model adaptation in text independent speaker verification. In: IEEE International Conference on Acoustics, Speech and Signal Processing, ICASSP 2008, pp. 4857–4860 (2008)

Young, S., Evermann, G., Gales, M., Hain, T., Kershaw, D., Liu, X.A., Moore, G., Odell, J., Ollason, D., Povey, D., Valtchev, V., Woodland, P.: The HTK Book Version 3.4. Cambridge University Press, Cambridge (2006)

Zavaliagkos, G., Schwartz, R., McDonough, J., Makhoul, J.: Adaptation algorithms for large scale HMM recognizers. In: EUROSPEECH 1995, pp. 1131–1135 (1995)

Zeidler, E.: Oxford Users' Guide to Mathematics. Oxford University Press, Oxford (2004)

Zhang, Z.-P., Furui, S., Ohtsuki, K.: On-line incremental speaker adaptation with automatic speaker change detection. In: IEEE International Conference of Acoustics, Speech, and Signal Processing, ICASSP 2000, vol. 2, pp. 961–964 (2000)

Zhou, B., Hansen, J.H.L.: Unsupervised audio stream segmentation and clustering via the Bayesian information criterion. In: International Conference on Spoken Language Processing, ICSLP 2000, vol. 3, pp. 714–717 (2000)

Zhu, D., Ma, B., Lee, K.-A., Leung, C.-C., Li, H.: MAP estimation of subspace transform for speaker recognition. In: INTERSPEECH 2010, pp. 1465–1468 (2010)

Zhu, X., Barras, C., Meignier, S., Gauvain, J.-L.: Combining speaker identification and BIC for speaker diarization. In: INTERSPEECH 2005, pp. 2441–2444 (2005)

Zochová, P., Radová, V.: Modified DISTBIC algorithm for speaker change detection. In: INTERSPEECH 2005, pp. 3073–3076 (2005)

Index

GPSR Compliance
The European Union's (EU) General Product Safety Regulation (GPSR) is a set of rules that requires consumer products to be safe and our obligations to ensure this.

If you have any concerns about our products, you can contact us on

ProductSafety@springernature.com

In case Publisher is established outside the EU, the EU authorized representative is:

Springer Nature Customer Service Center GmbH
Europaplatz 3
69115 Heidelberg, Germany

www.ingramcontent.com/pod-product-compliance
Ingram Content Group UK Ltd.
Pitfield, Milton Keynes, MK11 3LW, UK
UKHW022342280826
474907UK00003B/35

* 9 7 8 3 6 4 2 1 9 8 9 8 4 *